THE STONE RESTORATION HANDBOOK

A practical guide to the conservation repair of stone and masonry

THE STONE RESTORATION HANDBOOK

A practical guide to the conservation repair of stone and masonry

CHRIS DANIELS

THE CROWOOD PRESS

First published in 2015 by
The Crowood Press Ltd
Ramsbury, Marlborough
Wiltshire SN8 2HR

www.crowood.com

British Library Cataloguing-in-Publication Data
A catalogue record for this book is available from the British Library.

ISBN 978 1 84797 907 0

Acknowledgements
Thanks for help and support are given to Blossom, Lily, Rudi and Cadhla; Antony 'Hank' Denman; Richard Mortimer, David Good and students at Weymouth College; Ian Constantinides; Niall Finneran; Andrew Whittle, Ian Burgess and all other colleagues and past students.

Typeset by Servis Filmsetting Ltd, Stockport, Cheshire
Printed and bound in Singapore by Craft Print International

CONTENTS

INTRODUCTION

> Therefore when we build, let us think that we build for ever. Let it not be for present delight, nor for present use alone. Let it be such work as our descendants will thank us for, and let us think, as we lay stone on stone, that the time is to come when those stones will be held sacred because our hands have touched them, and that men will say as they look upon the labour and wrought substance of them, 'See! This our fathers did for us.' (John Ruskin)

Welcome. This is a book intended to give an introduction to the practical processes for the repair and conservation of stonework, covering techniques and application in the workshop and on site; it will also be useful to those involved in the commissioning and administrating work of this nature. 'Artisan' will be the term used for the person this book concerns itself with; my demand of you, the reader, is that you have an enquiring mind, an ability to learn and are willing to explore. The thread running through all the chapters will be the encouragement to enjoy and appreciate this work and objectives, in all their qualities and uniqueness. Complementing this will be an emphasis on the correct attitude to this work; the successful preservation of our built heritage is a task of such significance that, if you are not the masonry equivalent of a tree-hugger, then it may not be the career for you. Good luck with whatever you wish to do, or, better still, read this book, get excited, get involved, and perhaps join our ranks.

This replacement bracket in Bathstone involves a combination of topics covered in this book: bad practice – the building had been covered with a stone sealer to prevent damp, subsequently making the problem worse; decay – where the trapped salts have destroyed the stone and fabric as well as allowing moisture to be held in the fabric; stonemasonry – needed to work the replacement to match the original; fixing skill – to put it in; lime – for the mortar work and grouting of the core; and casting – copies were made in cast lime to put in as sacrificial units that would draw out the moisture.

OPPOSITE PAGE:

Exeter Cathedral, Devon. Built in 1153, this is a major Gothic cathedral that, like most buildings of this style and period, needs consistent maintenance and repair. This is where the skills of the conservator and stonemason become one.

The Essential Human Toolkit

We are all imbued with basically the same learning facilities and have evolved brains that can control our hands to

New bottom half of a heraldic shield, pieced in to replace the decayed section. Although the original was wholly made from the same stone at the same time, it shows that there can be extremes of performance from what would, technically, be described as the same material. (Photo: Harry Jonas)

perform wondrous feats far beyond the requirements of rudimentary DIY. If you want to do something badly enough, you can acquire the skills to do it. So the first thing to be pulled out of the 'toolkit' is *attitude*. With a positive attitude, the work, no matter how insignificant, becomes part of your life. To be able to do things that enhance, prolong and beautify the environment we live in is profoundly human and leads on to the next attribute – *pride*. We are not talking about sin here, rather the warm satisfaction in looking at what you have done, feeling that it was the best work you could produce and reckoning that it has been done correctly.

For those new to the use of tools, there will be the basic outlines for handling and using them. Rest assured that it is relatively easy and yet pleasantly satisfying to be dexterous in the use of your equipment. Just take care, practise in safety and if in doubt ask; most artisans are happy to help in the acquisition of skills by interested people, so knock on workshop doors and engage in conversation on site.

Added Interest

Throughout this book are photographs drawn from a variety of projects, many included to add another aspect beyond just the topic of the text. I hope they will begin to illustrate for you some of the range that can be found in work like this. Where there is little to illustrate in the text, interesting pictures will be shown that have some connection with what is being talked about. They will also demonstrate that many of the topics in the chapters are interrelated; in this work it is important to take a holistic view of issues and realize they cannot be categorized simply as having one thread. I have been practising this work for almost three decades now and have not had a single job that has not introduced new aspects, required novel solutions or needed the ability to adapt the technologies available.

The route to successful practice in this area of conservation work is not easily defined, for it is a *mélange* of many studies: architecture, art, history, science, technology and research, coupled with practical skills that, personally, range from welding and metal fabrication, through stonemasonry, sculpting, carpentry (not joinery though!), brickwork, plastering and mould making, to engineering and mechanics. Though this array may seem formidable, it is well within the grasp of most intelligent adults who find learning enjoyable and have the correct attitude for acquisition of new talents.

Sometimes complicated issues can be resolved with experience and common sense; the cleaning of delicate marble statues, rather than raising it to complex levels of chemicals and poultice, often just involves a bit of time and some warm water.

One of my first jobs as a conservation stonemason was to build a protective layer on this fallen chunk of masonry in Corfe Castle; it was to consolidate the exposed crumbling rubble infill in such a style as to still tell the story of what had happened by showing lines of rubble as they would have been during building. The use of flint has been suspended for some time for this work since then, as the realistic look was replaced with a safer, more anodyne 'rockery' style.

It would be nice to be able to decide on a proper definition for the work covered here, or for any practical work to the built heritage, and for ease it is termed 'conservation'; this term can be interpreted in many different ways, such as restoration, repair, renovation, maintenance, amongst others. The bottom line is that work carried out is appropriate to the needs of the building and its preservation; how much and to what, is up to those involved.

Checklist

In the majority of this work the aim is to preserve and protect the historic fabric fittingly, using the least aggressive techniques and minimizing the amount of new materials used; hence our most used term, 'minimum intervention'. Obviously, interpretation of levels of intervention deemed necessary for use in conserving historic buildings can vary widely; definitions become open to reason when taking the following topics into consideration:

- Original material used in construction should, if possible, be *preserved* as is, without moving, adding to or covering up with new material. Obviously, some change will be wrought on a building; necessary alteration/repairs (as will have happened in the past) are often to be considered part of the history.
- *History* is the visual and physical record of what has happened to the building and should be preserved as diligently as the original fabric. When the changes have been dramatic for change of use or expansion then this can have repercussions on the aesthetics.
- *Aesthetics* cover how a building looks, which is as important as how it is constructed, so work that is necessary for the preservation of the fabric should be appropriate in finish, and should not disrupt the design of the building or its detail by imposing suppositions and modern material finishes inappropriately.

Slightly false historic evidence of a cannon ball lodged in a wall lends interest to a building. The problem for us is how this should be treated.

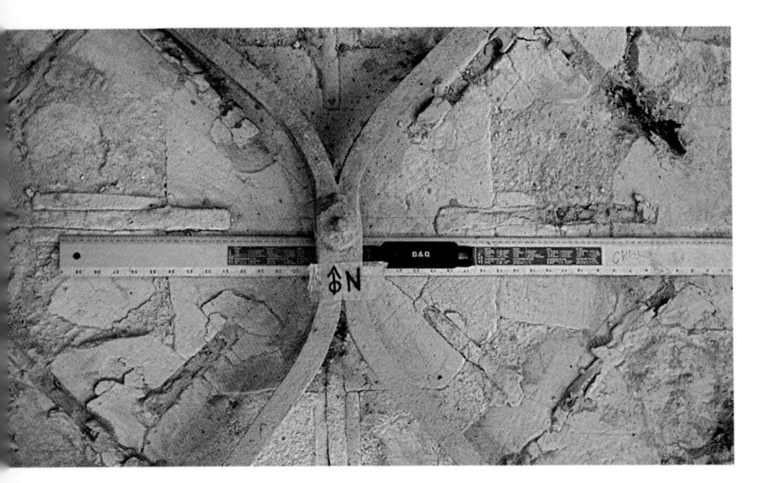

Wrought iron armature dating to the nineteenth century, used to support a medieval vault; this was subsequently replaced with a stainless steel frame to do the same job. Wrought iron is incredibly resistant to rusting; it was only the later ferrous fixings that had caused the problems.

- This work, if possible, should have a factor of *reversibility*. Our work should allow future historians and conservators, if necessary, to remove the modern interventions to get back to the original material, perhaps to carry out research or use more up-to-date repair techniques. Reversibility is often pushed down the table of desirable properties, especially if it affects the practicality.
- *Practicality*: this is paramount to getting the job done effectively using techniques appropriate to the task – cleaning large areas of masonry with cotton buds and litre bottles of de-ionized water is not practical! Thus a major factor here is common sense. The extent of the work, the use of specialist techniques and the level of intervention will rely often on the funding.
- *Funding*: clients are all too often given an indication of what a project will cost which, due to imprecise specifications written by non-practitioners, runs over budget; conservation specifications should be written by conservation professionals or specialist contractors to get the most pertinent deal, and these should deal with the building's needs.
- *Vulnerability and needs*: these aspects can come under threat so it is up to those with practical experience and skills to recognize the full extent of these criteria, how they can be worked with and how the repair work can be maintained.
- *Maintenance*: aside from necessary repair, the permanence of our built heritage relies on someone looking after it, so a schedule of required care should be drawn up for the structure and any intervention must fit into this programme; this should be viable in terms of practicality and economy to ensure it happens.

Salisbury Cathedral, like most buildings of this type, has permanent staff continuously repairing, replacing, renewing and caring for the fabric. Their work will never be finished, due to the small number of people involved, the size of the building and, most pertinently, the cost.

Chemical cleaning trials in Wakefield Cathedral undertaken as part of the investigation as how to revamp the interior. This work was carried out prior to our engagement and unfortunately the records were vague and did not help in any way; pressurized steam was our preferred method, which came up with comparable results that are more practical and kinder to the stone.

CHANGING TIMES

It is important to note that though the majority of conservation is carried out with the best of intentions, the use of seemingly correct materials in the wrong manner and application can not only be harmful but hasten decay. Conservation and restoration is an ever-evolving subject, and what may have been seen at a particular time as a suitable treatment, material or process may be acknowledged with hindsight, in some cases, to be detrimental. This may also come to be the view on methods in vogue today; while understanding of the factors involved in decay and how materials alters over time appears to be a much more precise science, there is still much mystery as to the precise nature and activity of many forms of decay.

Dichotomy

As there are policies (such as the Venice Charter) that adhere to the principle of producing work that should be readily identifiable as a modern intervention, it may be acceptable to use a stone or repair technique that may seem visually different, as long as it is physically compatible or its performance is supportive to the original material. This 'honest repair' is often evident in museum exhibits where it does not detract from the original piece, highlighting the found condition by not being afraid to show the extent of new work. John Ruskin was generally opposed to restoration that was hidden, in 1849 calling it '... a lie from beginning to end'. His opinion was that remedial work should be seen: 'Do not care about the unsightliness of the aid, better a crutch than a lost limb.'

We are all in this game for the good of the structure while simultaneously preserving our heritage. While work should be harmonious and complement the existing, we must appreciate that function begets form, rather than the other way round, the accepted approach demanded by many organizations involved in heritage preservation. Clients (generally those not in the field but paying for the work) are more inclined to invisible restoration, where the work carried out blends in to the original, presenting the patinated appearance of an untouched building, occasionally leading to renewal; these two approaches are often a cause of contention, and probably will always be so. This book is not overly concerned with the specific theoretical reasons for doing the work, or what is to be used for possible situations; all those issues are for the artisan and the other players to hammer out. Here, the wish is to give some idea of how to do it, and what kit is needed, though my opinions will occasionally surface, and helpful asides will add to the pot.

Mutual Understanding

The preponderance of educated people with few or no practical skills is an issue that affects all walks of life, but here we must concentrate on our world. There is a substantial chasm between those that pay for or need craftspeople's work and the actual craft workers themselves; the former may well feel detached from the structures for which they are responsible, and the latter may often be too busy actually getting on with the work to spend time carrying out the

The mortar design bench, for on-site production of a matched repair mix. Much of this work involves improvisation and adaption of methods to deal with singular issues; recipes will not be doled out in these pages as it could be dangerous to bring preconceptions to the project – they are all going to need their own tailor-made mixes.

necessary social milling. This will always be an issue; and overall the downside is a glut of experts and facilitators who contrive to organize, manage and inspect the work to our built heritage without real attachment. The society in which we find ourselves creates a system where people do not appreciate the experience and skills of the artisan, considering that to do manual tasks indicates a lack of intelligence and wit, and many artisans feel that with no practical skills these people lack the same, so there is a barrier between these two factions that has caused some ridiculous errors or wastes of time. This does not mean that all is lost. There are many people on both sides of the fence who respect the other's contributions; when this occurs the result is outstanding and worthwhile. The future of our profession should be full of cooperation and respect.

The Society for the Protection of Ancient Buildings (SPAB) was one of the founding bodies that concerned itself with conservation rather than restoration. In 1877 its founder William Morris wrote in the manifesto:

> It is for all these buildings, therefore, of all times and styles, that we plead, and call upon those who have to deal with them, to put Protection in the place of Restoration, to stave off decay by daily care ... to treat our ancient buildings as monuments of bygone art, created by bygone manners, that modern art cannot meddle with without destroying.

Sound words up to a point, that point being the last bit, where he underlines the reason why many academics and critics do not trust practitioners to do the right thing; the implication is that it is better they are not encouraged to do so in their mundane manner. We must set this to rights and prove him wrong.

During this conversion of a great house to a hotel, advice on behalf of statutory bodies was that all the mortar repairs be 'eroded' rather than follow the existing lines, as they wanted to show the eroded and interesting history of the building. This desire to produce a faux appearance of 'artistic' wear and tear can be a thorny issue between the artisan (who wishes to respect the building with work that states the quality of the hand) and the white-collar specialists who, in fear of being blamed for too strong an intervention, opt to reduce the result to a safe level that they can argue for. Happily this cowardly approach was successfully argued against and the sharp Victorian masonry was skilfully bought back to life.

The
WALCOT FRUIT & POTATO STORES

kindle
CLEARVIEW STOVES
kindle
High Efficiency
Cleanburn Multi-Fuel Stoves
01225 332722 0845 5050085
Visit our showroom
2 Sussex Place, Widcombe, BA2 4LA
www.kindlestoves.co.uk
kindle
kindle
Low Cost Finance
Now Available
0% APR
Subject to status and terms

CHAPTER ONE

DECAY OF STONE

The problem with buildings, and the reason for this book, is that over time they will wear out. The process can pass unnoticed until it gets really bad, as buildings are at the mercy of the ravages of time, the elements and, quite often, of people. With such a hardy material as stone it is often an accumulation of factors that will set about the demise of a building, and once started it is very difficult to reverse the process.

Heritage conservation has the responsibility for reducing the impact and, where possible, setting things right – but always without losing the essential character of the structure.

Roman column, Nîmes, France. Gently wearing away with time, inevitable but true to nature.

What is the Problem?

The first step is to identify what is happening and what is causing it. A building will tell us what is happening to it; interpreting the story requires an ability to read the signs, plus common sense and intelligence. What will become noticeable in this chapter is that there is one unifying factor: water.

The Inspection

The extent of investigation here is assumed to be an overall view of what is happening to a building; the depth of this

OPPOSITE PAGE:
A rainy day in Bath. The façade of this shop is soiled by scurf, the result of acid rain pollution and accumulation of sooty particulates. The impervious painted sign is causing water to run off and keeping the stone clean directly below it by washing, while under the cornice the scurf builds up.

THE ROLE OF WATER

It can be generally assumed that where there is decay of stone that is not wholly caused by structural movement, water will be or has been significant to the process. Stone will at some time come into contact with water, or be contaminated by moisture; problems arise when the water carries soluble materials, is blocked in its travel or exerts unwanted pressure. If a stone could be isolated from any form of moisture it would probably remain whole, with no significant degradation taking place. As this situation is one that is impossible to maintain, because buildings are exposed to the elements, it is in the best interests to identify the deteriogens, to render them and their source inert or to remove them. Obviously artefacts such as monuments or statuary can be insulated to a degree from moisture ingress by the use of DPMs underneath or where they touch walls, but this is impossible with buildings.

SIZE MATTERS

There is a physical action whereby water will move through very fine spaces more quickly and easily than through large openings; this is known as capillary attraction. Buildings can often have myriads of fine cracks, especially in cementitious materials, which physically draw water into the fabric; so it behoves the inspector to get up close and check these out.

Wall monument, the 'before' picture showing the effect of decay from salts and rusting iron dowels. After the repair work, plaster was removed around and it was spaced out from the wall to prevent moisture travel contamination. Note that moisture *travel* is just that, while moisture *movement* is the result of disruption or expansion; try not to get them mixed up.

A dome cover stone has been lifted and the poor condition of the fabric discovered; decades of damp trapped inside by misguided and budget attempts to prevent this had caused the building to rot from the inside outwards.

study will be defined by what is found. First, make sure you have all the kit needed and try to choose a good day: too much sunlight may mean that significant details may be hidden in shadow (though you will get nice pictures) and, obviously, it should not be too wet.

Ground Level

Start at the bottom and have a walk round. Ground water is a major problem so you will be looking to find a soil level lower than the floor level; if it is not, then moisture can creep into the masonry and travel upwards before it starts to exit. Is there pooling water at the base; does the terrain slope down to the building?

The area first affected is that up to a metre above ground level; evidence will be denuded pointing and decaying stone at the evaporation zone on the outside and damp decaying plaster inside. Often the previous attempts to solve this problem (which is only the symptom) will be hard (cementitious) pointing or impervious plasters on the inside; these 'remedies' have the unfortunate effect of pushing the moisture higher or into other areas, often leading to exacerbated decay of timber in the structure and collapse of internal floors or roofs.

Good drainage is the key here so figure out why the water is choosing to enter the building rather than drain away in the ground; often it is previous work that is at fault. Concrete aprons around churches are a typical example of attempted remedial work. Designed to form a wash or trough system to divert the water, they usually shrink or crack, allowing water in but preventing it from evaporating; and thus the crux of all work to historic buildings (and obviously modern ones as well) is highlighted: moisture will always get in – it is letting it out that is crucial, hence our promotion of porous repair.

Surface water drains should be cleared regularly.

The evaporation zone is the area of the wall where moisture rises through the fabric to a point where it cannot go any further under its own pressure and decides to get out. Here all the soluble salts trapped in the surface of the stone have gone through a series of wetting and drying cycles, promoting crystal growth that exerts pressure enough to break the bonds of the grains in the stone, thus leading to devastating erosion. Note that areas of decay in this manner tend to be worse in detailed parts of the building, such as jambs, mullions, quoins, etc., where the surface area of the feature is high when compared to that of ashlar. This allows faster evaporation that causes more moisture to travel to these regions, and so the problem grows.

Rising damp exhibits itself in the deeper colour of the limewash, while the excessive moisture is promoting the growth of algae at the foot of the wall.

A bitumen apron (or cement equivalent) has often been used to prevent water collecting at the foot of the wall. Unfortunately here it has shrunk away from the stone allowing water to run easily into and under it, where it will be trapped unless it rises through the masonry; thus the perceived cure has made the problem worse.

Here a tomb restoration was carried out under some ambiguous specification, with the slabs in the enclosure being relaid; it was only after rainfall that the lack of drainage became evident as a potential problem.

Off the Wall

The next area is the masonry itself. Well-designed stonework has detailing that effectively disperses or reduces the impact of water; this includes projections of cills with drips, string courses, hood moulds and sloping (weathered) surfaces, all designed to throw off or keep water from running into the fabric. Check the condition, looking for open joints, mortar breakdown, fractures or damaged/missing elements. Pointing can look moth-eaten but still be doing its job; if it is near the surface of the wall and cannot be blown out, perhaps leave it until next time?

The colour of the stonework can help here, as damp stone is deeper in colour than dry. Increased moisture levels will also allow biocolonization, starting with green staining from algae and possibly leading up to woody plants rooting in the joints. Wet masonry will also promote fungal growth in timber that is set in the walls, as well as making damp environments ideal for wood-boring insect attack.

The Roof

Inspection of a roof can be difficult without good access, but a visual check can determine missing slates or tiles; broken pieces of these can often be found at the base of the building, prompting further investigation.

Ridge tiles on old buildings become dislodged as roof timbers settle, with the pointing and bedding mortar breaking out of the joints. As water ingresses it will cause even more movement in the timber and thus the problem feeds itself.

Lead flashings and valleys should be checked for tears, holes or missing pieces. Pointing on flashings is prone to failure as the lead expands, or where thin slivers of the mortar become detached.

Plants can slowly and surely inveigle their way into the finest cracks and grow to disrupt the best masonry; an early snip with secateurs would have prevented this.

Torre Abbey, Devon. New lime render shows the recent work carried out here. The corner is a humid environment and promotes the greening up of the roof stones.

Albert Memorial, London. This had internal guttering but the tanks behind the mosaic panels swelled and leaked water into the structure. Under its heavy lead sheeting cover, this is an iron-framed monument, which, once it started rusting, needed huge amounts of remedial work.

Coutances Cathedral, Normandy. Church roofs are often intimately constructed with the masonry so the building must be treated holistically. The erosion to the stone in the foreground is from ropes used to haul stone up during construction.

When inserting or repairing flashings, a good application of waterproof mastic to the back of the joint will help; rather than setting the sheet in with mortar, use lead wool (or shredded lead) beaten into the joint instead.

Anywhere where mortar is attempting to seal or join disparate materials (stone to lead, and so on) will be prone to failure as the differential expansion rates of the materials, combined with a lack of adherence, can create hairline fractures.

Obviously all gutters and downpipes should be clear of debris and not leaking; the wall behind downpipes is so often a place where damp problems originate, so check for water patching and greenery behind these.

Salts

It is accepted that old buildings have to 'breathe', though a more precise term would be 'perspire', and to obstruct this action or stop this happening is a bad thing. It is in the course of this attempt by moisture, present and mobile in a building, to attain a state of equilibrium that some factors of decay can come about.

Stone has powdered off through soluble salt decay, and a small mortar patch has been applied to consolidate and provide sacrificial material.

When the air surrounding or inside a building is warmer, drier or of a lower pressure (such as when the wind is blowing), moisture that is present in the wall will attempt to leave by evaporation through a suitably porous place in the fabric. If the moisture has passed through or been contaminated by material containing minerals that are soluble in water, these minerals will be carried in solution to the place where the moisture evaporates. Once the moisture has evaporated to a point where it is all gone or the concentration of the minerals is too great to stay in solution (saturation point), then the minerals will be deposited on or near the surface of the stone.

Hidden Salts

Soluble salts are carried in moisture and these will, through drying and wetting stages, cause crypto-florescence where the moisture evaporates out of the masonry leaving the salt to turn into crystals. The (micro-) crystallization will, over a period of time, bring about enough pressure to break down the matrix of the stone, with subsequent loss of material.

The first indication of salt is efflorescence on the surface. This is a white dusty powder that can be easily brushed off. Also look for the absence of biological staining; as the majority of these salts are inimical to plants, they can effectively prevent the build-up of such. Be aware, though, that the runoff from metal, such as window grilles, ironwork or cuprous material, will give the same effect.

Fresh Powder

The next stage is that the surface becomes friable, powdering off to greater depths as the loose material becomes detached. This should be happening in the sacrificial elements of the building (mortar and plaster), which can be removed and reapplied, providing a fresh defence against the destructive forces of the salts.

When impermeable surface coatings, such as cementitious mortars or waterproof paints are used, the salts in solute materialize at a place where the water can evaporate out of the wall. This will be in cracks in the render, around openings and timber, and on the interior surfaces of the building. If the wall has cementitious pointing it may be seen that the stone edges abutting the pointing will degrade, leaving a trellis of cement mortar almost 'floating' on the surface.

The same thing may happen when micro-porous surface treatment (and here we are talking about all weatherproofing finishes and 'permeable' paint systems) is carried out but does not have a pore system large enough to let large molecules of salt pass through, or when the surface pores of a stone become blocked by the intrusion of some other mineral, as in contour scaling in sandstone.

Carved label stop has a Victorian surface finish (this is easily recognized by the orange/tan hue to the stone), through which moisture is trying to evaporate. It is pranked with salt crystals growing out of pinprick scars that will slowly come to cover the whole surface and eventually all detail will be lost.

Contour scaling of sandstone occurs as the face delaminates in layers unrelated to the bedding.

Bird droppings are not only hazardous to health, but will deposit minerals on the stone and provide a nutrient-rich environment for plant growth.

Bunhill Fields Burial Ground, London. This shows the visual effect of acid rain on sandstone and limestone; where the water hits the surface on the sandstone (left), soiling from reaction is overall. On limestone (right) it is basically the reverse, with water-washed areas clean and sheltered bits staining up.

Humidity

Some salts are hygroscopic; that is, they tend to absorb moisture from the air. The result is most noticeable when there are high levels of humidity present inside the building, the prime example being when a building has public access and has central heating installed. This causes uneven periods of drying, warming and cooling of the building that bears no relation to the weather outside.

The product of many people breathing, particularly in a warm atmosphere, will be high levels of humidity. Stone will usually be several degrees cooler and so the moisture will tend to condense on its surface, resulting in a wet stone that will have salts (that are present in the stone) in solution near the surface. Another example of this would be mullions: weather variations produce salt concentration on the interior surfaces, which leads to their rapid decay. The same can be seen where the building acts as a condensation point for mist and fog; this often occurs in isolated positions or if the building is near water.

We will look at the effect of plant and animal life later, but it is worth mentioning the effects of bird droppings, which can produce a localized source of humidity.

Pollution

Soluble salts also play a role in the decay of stone through contamination from airborne pollution. With the burning of fuel and materials on a large scale for energy production has come the infamous phenomenon of 'acid rain'; though this has been around for many years, it is the increase in industrialization that has caused it to rise to critical levels.

Modern contributors to airborne pollution are vehicle emissions (diesel being the worst offender) and extensive use of fossil fuels for heating and power. Stone is affected by airborne pollution in various ways chiefly according to its mineralogy.

Acid Rain

Sulphur dioxide, nitrogen oxide and, to some extent, carbon dioxide (all products of burning) when in rain will form acids that can react with minerals in stone, most actively in limestone where the (sulphurous) acid changes the calcium binder into calcium sulphate (gypsum).

Sitting on a garden wall in Stoke sub Hamdon, this lovely model of the church is made from tiny blocks of real stone; sadly decay has no respect for the efforts of man – time for some repair!

DESALINATION

The removal of salts (desalination) would seem the obvious solution. The general process is quite simple: spray the area concerned, saturating it with de-ionized water (DIW), then lay a poultice of absorbent material (acid-free tissue, sepiolite or paper pulp) soaked in DIW onto the stone, leaving it to dry so that the salts are drawn into the poultice. Remove the poultice, place it in a container of DIW, and test for electrical conductivity; high salt concentration = high conductivity. Carry out laboratory tests to identify the salts.

Trinity College, Cambridge. Desalination of clunch using acid-free tissue, prior to repair work. The poultice is sprayed with de-ionized water (DIW) and allowed to dry. Later placed in more DIW, its conductivity is measured, the process being repeated until a significant drop is reached.

ISSUES

As the poultice dries and shrinks off the wall, it stops absorbing (perhaps a case could be made for moist mediums or removal before drying). How the pore size of the stone affects the salt transfer is not being addressed here with the use of universal mediums. The problem can recur if the source of the salts is still present, or if there is salt in the masonry that is not removed, as the resulting low concentration of salt will encourage further salt migration.

There are so many parameters that can alter the effects that it is almost impossible to say whether this will be successful, so beware of relying on a single concentration reading. There is no guideline on how much salt is safe and to draw conclusions from a single sample is not sensible; limitations should be recognized. There is research all over the world on this subject, and while they have wildly different approaches and technology, they tend to agree on one thing: there is no telling what will happen when trying to desalinate porous building materials affected by soluble salt.

Complete desalination can only be 100 per cent certain if the stone can be completely isolated from any further moisture travel before, during and after the process; this is patently impossible for a building. After desalination it may be necessary to modify the environmental conditions to prevent or hinder further outbreaks.

Trinity College, Cambridge. After desalination of the window surrounds and mortar repair of decayed stone, a ring beam was inserted along the length of the range, which involved pouring tons of OPC concrete laden with soluble salt into the fabric; arguments for the use of benign lime concrete to be used instead were dismissed.

Scurf builds up on a typical limestone window. Note the ugly cement repair to the mullion.

Scurf

Being soluble in water, the newly formed material washes out of the stone and crystallizes as a gypsum crust over the surface, hardening in air to form a protective barrier while the stone below decays even more. In cities this barrier material becomes blackened from sooty deposits to form scurf, identifiable by its location in sheltered areas of the façade; conversely where it is rain-washed the cleansing action of water prevents a build-up. The difference in thermal expansion between this gypsum crust and the limestone will eventually result in loss of all surface detail, as the scurf breaks free of the stone.

Magnesium (Dolomitic) limestone can be really badly affected by this action and will decay to a greater depth, termed cavernous decay.

The bronze plate set in this limestone shield (shown here lowered to the horizontal for attention) is producing biocidal wash that prevents algal colonization directly below it, while life clusters at the edges.

Salisbury Cathedral, Wiltshire. Built of calcareous sandstone from Chilmark, the structure has had serious trouble with scurf causing loss of the detail; this is unusual considering the low industrial pollution of the region.

Sandstone gravestone showing staining and delamination. The green areas are lime mortar fillets I put in about fifteen years ago that have become little allotments of algal growth.

SMOG

Urban pollution in London, due to the burning of the local high-sulphur coal, caused the first legislation on smoke pollution; Edward I attempted to curb smoke in 1273, and transgressors who burnt coal in the city could, like one London manufacturer in 1306, be tried and executed. In 1578 Elizabeth I refused to enter London because of the smog and by the beginning of the eighteenth century the deleterious effects of pollution were affecting buildings, vegetation and even clothes in every town of any size in Britain. In 1661 the Ballad of Gresham College describes how the smoke 'does our lungs and spirits choke, our hanging spoil, and rust our iron'.

Smog is a portmanteau of smoke and fog and on occasions has been so dense that it was nicknamed 'pea-souper'. In 1952 there was a 'pea-souper' that lasted for four days in London and killed 12,000 people; this led to the 1956 Clean Air Act and the subsequent cleaning of buildings in the capital.

Stains

Sandstone, on the other hand, is composed mainly of quartz and feldspar, which are not reactive to acid; it will show discoloration of the rain-washed surface where the acid oxidizes other minerals (usually iron) in the matrix. Using an alkaline cleaner on this will make the problem worse; always test first. Knowing these facts about staining will help in identifying the stone type in buildings; the effect is markedly varied, some stones being almost impervious to the changes signalled by staining while others can break down very rapidly (in building terms).

Metal

The oxidation of metal (rusting in steel/iron and verdigris on cuprous metals) is enhanced when the moisture is acidic, so pollution will affect these materials on buildings.

Expansion

Ferrous metals will expand prodigiously as the rusting takes place, subsequently lifting or breaking stones when such metals are used as cramps or fixings. Diligent builders in past times were aware of this and would attempt to seal off the metal by encasing it in lead; when done well this was very effective, but if it fails the results are the same as if they had never bothered.

Expanding metal is pushing this vault apart. The plug in the centre of the boss-stone is attached to a huge wrought iron brace above, inserted to hold up the vault; the pin tying it back was causing more damage as it rusted. The brace was subsequently replaced with a custom stainless armature.

There are seven fractures along the corners of these blocks, showing where cramps are rusting and expanding.

Widening of bedding joints becomes apparent as cramps lift the courses above (one single cramp has the expansive strength to lift many tonnes of stone). The mortar will become loose and let more moisture in – exacerbating the problem.

If the metal (cramp) expands sideways it will blow out the corners of the stone. The first indication may be a hairline crack running diagonally across the corner, or split mullions and jambs.

Pintons, *ferramenta*, railings and brackets have usually had their lugs set into lead in the stone, which is good, but if the paint on the exposed metalwork is not kept in good condition then the result will still be rusting, with increased pressure from the solid plug expanding directly within the stone. Wherever possible, ferrous metal should be removed and replaced with a suitable alternative.

Gate pinton has been set in lead (visible to the right), but the neglected paint has allowed it to start rusting and expanding in the jamb-stone.

Window *ferramenta* removed, showing the relics of rusting tips (the missing material has expanded and flaked off) that were replaced with welded stainless steel lugs.

Staining

Any metal fixtures on a building will be subjected to the weather, mainly acid rain, which often causes oxidization at the surface. The resulting altered metal ions become detached, entering into solution and are washed onto the stone below; this stone, if dry, will absorb the solution leaving the metal particles in the surface layer. The process of chemical alteration continues on this layer and after time the stone is stained the colour of the oxidized metal – brown for iron and green for copper. As mentioned, the first indicator of this happening is the biocidal action of the sulphate killing any biocolonies in the wash area.

Cleaning sandstone with alkali- or acid-based solutions can mobilize metallic ions in the stone, moving them to the surface where a whole new set of problems can occur.

Verdigris staining to the plinth of a bronze statue; the main concentration is on the sheltered part of the stone.

> **WIRE BRUSH**
>
> The temptation often arises to clean the surface of stone with a wire brush – don't! Discounting the abrasive action that can leave myriad scratches across the face, eventually leading to strange patterns of discoloration, consider that wire brushes wear down – so where do the old bits go? The tiny bits of steel end up embedded in the stone and as time goes by they go rusty and show up.

Other Problems

Ice Jack

The interaction of metal and salt is not the only expansion issue for stone; ice can be incredibly damaging to masonry. The process is very simple: if the temperature is low enough to freeze water, then the volume of ice formed will be ten per cent greater than the original volume of water, thus outward pressure in the stone will cause it to spall off. This affects copings where they are natural-bedded and the water seeps into the interfaces between layers; it cannot really be stopped unless a (lead) flashing is installed.

Fine-jointed structures may accrue moisture in the joints which, when freezing, can jack the stones out of position. Over time this can cause significant movement; it also allows more moisture to sit in the joints and attack any metal inside.

Pore size is one of the factors that control the susceptibility of stone to frost damage; it is the type and disposition of the pores that are the important factors. Stone with a low saturation coefficient will usually be able to resist this process. Large open-pored stone is often more resistant to freeze damage.

An ingenious solution to ice jacking found during work on the Legislature in Winnipeg, where temperatures can drop below minus 40°C. Knowing that putting a mortar joint under this exposed small stone atop of a statue would be a waste of time, the mason cut out a rebate in both stones and put a block in to locate the top stone. With no mortar in it, the joint can fill with water, freeze and expand but the loose block prevents the stone coming adrift; a fine example of not trying to beat the elements, but working with them – figuratively bending in the wind.

Ice cracking of Portland stone, due to moisture in fine vents freezing when wind with a minus 20°C chill factor occurred – a rare event in the UK, but with imminent climate change perhaps a sign of things to come.

The slow growth of most plants can make their encroachment into masonry structures a bit of a surprise in some cases; this probably caught everyone off guard hidden away in Bunhill Fields Burial Ground.

Plant Life

The surface of stone is ideally suited for colonization by lower forms of plant life; indeed the patination of lichen and moss can enhance the appearance of an old building – to a point. Some of these are evolved to nourish themselves from the minerals in masonry or from the products of pollution; they in turn provide nourishment for other species – life survives and flourishes in seemingly barren environments.

Occasionally the seeds of higher plants will get trapped in the cracks or the vegetation and germinate; with the moist environment and the nourishment provided by the breakdown of other vegetation they can thrive. Roots and tendrils spread into joints and fissures causing expansive jacking and fracturing, as well as creating and transporting a moisture-rich environment that encourages further growth and colonization. Such plants are also good indicators of overdamp masonry, leaking water or the presence of animal life.

Animal Life

The most obvious are burrowing creatures that will cause voids in the core, leading to possible collapse; check for holes and spoor. Masonry bees will bore into soft mortars to make nests; spotting their holes is easy in larger joints.

Wood-boring beetles can completely destroy the structure of lintels and other timber in walls. As the masonry is usually built onto or around this, the damage can be substantial. Check for exit holes of the adult beetle; once again, moisture in the fabric will help these pests to propagate.

Masonry bees boring into lime mortar on the back of my house are not a good thing to leave – buzzing, stinging insects do not make for restful nights, so I will need to sort this out soon!

Soft propping to a sixteenth-century ornate plaster ceiling in Holme Lacy, Herefordshire. This was a project that had a variety of problems but one simple cause: a roofer neglected to put in a lead sheet when building a dormer, which let water into the roof. The fully boxed-in rafter, having no ventilation, produced an environment moist enough to allow attack by beetle and fungi, subsequently completely rotting away. The weight of the slate roof then transferred to the tie-beams of the ceiling and caused a deflection enough for the plaster to fracture. The ensuing repair and consolidation of the ceiling cost over £75,000; the cost of the missing lead was about £5.

Human Issues

The first problem is that at any point in the past, builders may have cut corners, used incorrect materials or been insufficiently skilled to carry out work to a decent standard; often it is a combination of all three, so these need to be considered together.

Out of Bed

The precept that stone should be natural-bedded is actually a simplification; the principle is that the bedding of a stone should be normal to the force exerted on it, though exceptions exist. For example, stresses would be too much for a markedly bedded stone made into items such as mullions and jambs, so they tend to be end-bedded, but as this can lead to vertical delamination they often need repair work.

Cornices and copings with drips and fine moulding detail, such as dentils and modillions, may lose these, as the decay processes do not have too far to travel through the layers before allowing detachment – there have been many injuries from falling chunks of masonry because of this. These features should be laid on edge if possible, especially if there is a chance of delamination. Gargoyles and projecting details will also suffer from this; it is down to the designer or mason to overcome this problem when producing the feature.

Hamhill stone blocks used for chimney stacks have completely broken up due to the lines of weakness from clay bedding in the stone itself.

Lintels can fail when settlement occurs or the loading is increased beyond their capability; fractures need to be scissor-pinned and, if necessary, a steel strut inserted to help support.

Obviously, similarly set stones must be used for replacement if this is the only way, though common sense will dictate that measures to aid the stones' longevity be taken as appropriate.

Suitable stone is always an issue. Here in Shaldon, Devon, the church is built of various stones with all the dressed detail in polyphant stone (quartz porphyry, known as elvan in Cornwall), a beautiful stone for carved detail and finish, but hopelessly ill-suited to outdoor use, especially near the sea.

Base Matters

Foundations are nowadays designed to take and spread the load of a building, using a considerable knowledge base of science and maths. Going back into the history of construction it was often the case that walls were built directly onto the earth, with minimum or no load allowance.

'FLOATING' CATHEDRALS

Salisbury Cathedral soars to over 120 metres and is built on marshland; the foundations for this massive structure are only 1.2 metres deep.

Winchester Cathedral, also built on marshland, was under threat of collapse in the early twentieth century due to the insubstantial base on which it was built. This was so waterlogged that the repair was carried out by a professional diver who took five years, working totally submerged in muddy water, to place almost 30,000 bags of cement and 115,000 concrete blocks under the building.

Settlement

All buildings settle as the loads are distributed throughout the structure: the art is to ensure this happens evenly. Historic buildings were often built on an ad hoc basis using whatever materials and skills were available in the locale; this means that there are often very visible signs that the building has moved. If this has not caused undue stress to the point of fracturing stones and it can be seen that everything is where it is going to stay, then it should not be an issue.

If there are signs of continuing movement, such as recent pointing not being solid in joints, clean surfaces on breaks, then checks should be run to ascertain the extent and timing of the issue.

Mortar Loss

Fine ashlar work can be spoilt when joints are denuded of mortar and/or there is some movement that causes the faces to touch, with subsequent spalling and fracturing. Mortar is essential to even out loading over the beds of stone, and if pliable enough it will allow some movement to take place without causing stress points. Using a mortar that is too hard and brittle, or one that adheres strongly to the stone, can lead to stresses and fracturing due to its inability to give.

Grouting the core of the wall. Note that the filling starts from the bottom, with outlet pipes for the grout in the course above to show a successful fill; once it reaches these they are plugged with golf tees.

Double-skin walls that have a rubble infill at the core can sometimes slump as the (slackly mortared) loose particles settle or become washed down with water ingress. This can lead to bulging and collapse as the wall skin is pushed away and loses coherence. Grouting is a process of filling the core of the wall with a liquid mortar that will solidify and bind the stones in place. It can also be used to fill walls that have unwanted voids or cracks, and thus create a solid structure.

WALL FAILURE

Roof timbers may start to bend under the weight of the tiles, or wall plates may rot and break up. In either case the result may be that walls are pushed out at the top; the same applies to arches and vaulting as they settle or become overloaded. This may cause individual stones to rupture or split, which increases the movement potential.

Adequate ties, replacement timber and hidden tie- or ring-beams should be considered as remedies. Many historic buildings already have ties across them or metal or wooden beams set in the walls and if these fail they can lead to movement. Buttressing to the outside, as widely used in gothic architecture and much vernacular work, is simplistic and effective but not used very often today.

Buttress to prop a corner of church in Bridport, Dorset. Much of the stone surface has been repaired with a textureless lime mortar that is failing to weather in any way sympathetic to the stone.

CASE STUDY: FRESH AIR

Many of the problems with the internal degradation of historic buildings can be put down to changes in their use; this often results in modifications to make them comfortable for people used to dry, warm habitats, so the natural or accidental ventilation that prevents the build-up of damp is stopped. Many solutions have been produced to make ventilation that is effective while still blending in with historic buildings.

An old folly built as a bathhouse had been converted to a dwelling. The problem was that it was not designed to be weatherproof as it had no windows or DPC and it was built above a water course. Over the years many attempts to get it dry had all accumulated to make the problem much worse: the outside had been treated with a waterproof silicone coating; inside, the domed roof had been rendered with a swimming pool cement; and hard plaster had been applied to the interior walls. Water had accumulated up to four metres high in the fabric and when the plaster was removed the brick core was found to be rotted out, soaking and spongy. The brick was subsequently replastered with porous lime mortar and bespoke vents put in below the windows.

Water leaking out of the fabric on top of the window surround.

New plastic tubes lead through the wall to the outside.

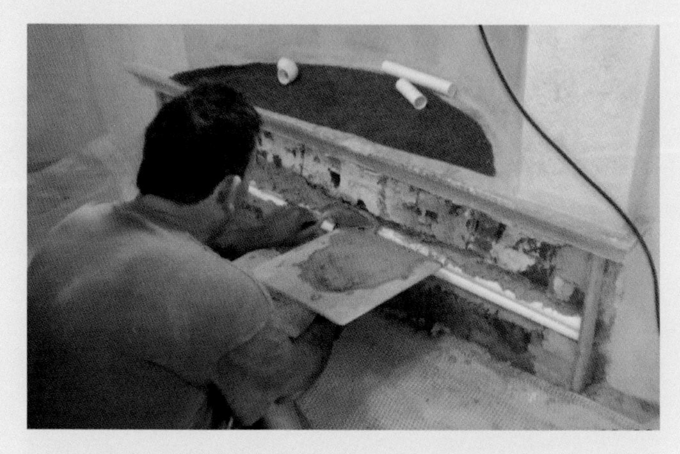

Outfall pipes are connected to internal pipes leading to the sides of the plinth; the latter are plastered over.

Interior holes for the venting are left exposed temporarily. The mortar is applied by spray to reduce shrinkage.

Drilled stone caps to blend in with the building cover the ends of the pipes; the base coat is scored for the top coat of plaster.

These were designed for the outside of a historic building to blend in with detailed stonework; each has an adjustable vent plate, with a stop screw to prevent it spinning.

Evaporation stones for the outside of the dome allow moisture to soak into their backs, aided by slots, and the large holes create a drying swirl to help the removal of excess moisture in the fabric.

CHAPTER TWO

PRODUCING A REPORT AND SELECTING THE STONE

The Conservation Report

It is a general requirement that any work to the built or movable heritage should be accompanied by full documentation of the project, including a conservation report; this is essential for any grant-aided work. This means that for the sole contractor running their own job it is necessary to visit and examine the structure, produce findings, recommend treatment, and comment on the results – quite a lot of work, so it pays to be on top of the game.

The Site Visit

This will be the time to assess the task, cost the extent and get the necessary paperwork started, so make it a fruitful experience. If it is an established contract and you are a subcontractor, then it will be necessary to be inducted by the main contractor, or have a risk assessment in place. Visitors to any site are expected to have all PPE and suitable clothing; overalls and stout site-only boots will make cleaning the car less of a chore.

Consider taking a partner if there is a need for ladders or if the site is deserted or dangerous; a mobile phone in your pocket and someone waiting for you to return can be lifesavers.

OPPOSITE PAGE:
Palace of Westminster, London. In the mid-twentieth century, the Canadian Prime Minister Mackenzie King collected several reclaimed statues from London masonry yards and transported them to Canada. To establish their provenance I had to search for the original sites, which included a survey of the statuary of the Houses of Parliament, to match the style and materials.

Every picture tells a story; this is Elton House, Bath. This Georgian house, nestled in a historic square, presents a sorry sight. The soiling of the surface is from pollution that has been rife in this city, causing all the buildings to be almost black by the mid-twentieth century; a programme of cleaning was instigated and brought most back to the loveliness we see elsewhere in the streets, but not here. The white patches below the cornice and cills are where the build-up of scurf has broken away, exposing raw powdering stone beneath. (By contrast, the pub to the side is well presented, cleaned and with a limewash to protect the stone.) We used to lodge in this house while working in the city; we carried out small internal repairs and put new lead over the bay window to pay for our keep, but sadly the façade has never received the care it needed.

Wearing a climbing harness is now compulsory for high-level work and survey, here used while carrying out routine maintenance to the King's Statue in Weymouth.

Forewarned

Many historic buildings have printed information about them, so gather the relevant details; with the internet it is simple and effective to find out as much as usefully possible before visiting. Make a note of construction dates, style(s), significant architects and owners; when you meet the client they will appreciate your professionalism and interest, while the information can be helpful when viewing and taking notes.

DEAD ZONE

A qualified climber and 'conservation' expert, while conducting an inspection of a tall stone industrial chimney was lowered down into the interior of the structure and almost died. The reason was that there were no openings in the depths and the higher level of carbon dioxide caused 'foul air' to build up in sufficient amounts to induce unconsciousness almost immediately. Luckily the law requires climbers (or those using rope access) to never work alone, so the hapless individual was rescued before he died.

Workers in historic buildings can encounter bizarre hazards; here an understanding of basic science would have prevented this near calamity.

First Look

With a good-quality camera to hand or, better still, on a strap around your neck, stroll around the building taking high-resolution pictures of the façades and areas of interest; these are useful for placing elements in context and for checking afterwards for missed items. Binoculars used to be widely used in this process but now taking a high-resolution picture allows close inspection of distant elements, and can be annotated immediately on a tablet if needed.

Mobile phones and cameras are fairly delicate items, which can be easily damaged on site, dropped from a height (neck strap for camera again!), soaked by bad weather or clogged up by dust; get protective cases for them and keep in a dry place when not being used.

This drawing by Dimitri was part of a lesson to get the students really looking at a building in terms of how it all works together in its proportions and elements. Much architectural work is based on simple shapes; here there are squares and rectangles that repeat, giving uniform proportions that please the eye. It is worth getting a sketchbook out and finding a building akin to this and discovering how it all works (for example, where the arches start). See how its component parts create a whole that has good proportion, and do try to learn all the architectural terms.

Divide the structure into manageable sections and title them in a notebook; if necessary make diagrammatic sketches that can be linked to photographs. Starting at the bottom left-hand corner of each façade, number the courses and elements in horizontal sweeps – this is standard practice for masonry drawings, and very useful for discussing with other involved parties. Number all notes, have abbreviations (MR = mortar repair, etc.) to speed up the process, and express decay in percentages of each stone or section.

If it is raining or damp then have a voice recorder as an alternative to a notebook, but keep this organized – describe the section fully and relate each back to your photographs.

Close Up

Set up a rule or photographer's tag next to the elements to be photographed (attaching them with gaffer tape if appropriate), and take clear natural light shots, preferably without flash, for reference. My personal choice is a 17mm pancake lens with a clear filter on (to protect the glass and enable easy cleaning/replacement); this will take good close-ups and fairly wide shots (essential if on a narrow scaffold or a ladder) without needing to fiddle about when your hands are dusty.

For eroded detail try using raking light from a strong torch or remote flash held close to the surface; this can reveal a surprising quantity of information. With rules and setsquares measure the outer dimensions of jambs, string courses, cornices etc., then break these down into individual mouldings. Allow for erosion when taking measurements; usually internal mitres will be the best markers. It is often found that they have been built up using whole numbers to make production simpler; this allows intelligent extrapolation of sizes to be made where detail is missing. Profile gauges can take off mouldings or, if possible, slide a piece of plastic paper into an open(ed) joint and trace the true shape; do not use lesbian curves here, as they will often distort, unless the situation is appropriate.

Wear the Right Kit

Protective gloves are necessary for any investigation that involves dismantling or rooting around in dark spaces, coupled with a head- (helmet-) mounted light rather than a hand-held torch – if a torch is used, have it on a wrist strap. A sturdy shoulder bag with pockets will keep tools and equipment to hand at all times, storing samples safely as well as allowing greater freedom of movement.

Kneeling on scaffold boards or bits of masonry is extremely painful and can cause long-term damage, even if it does not feel too bad at the time; use kneepads for any prolonged low-level work, and also to protect trousers.

Sampling

If samples of materials need to be collected, use appropriate-sized containers and a permanent marker pen to reference them. Stick a square of masking tape next to the point of collection and write on the reference number, then take a picture or mark on your drawing.

When looking at the particles in stone/mortar carry a Geology Grain Size Card – plastic and the size of a credit card – handy to note the physical properties of the material. Be logical when sampling, and get typical examples from unobtrusive locations; internal mitres tend to be the most protected and are good for paint and mortar finishes, but remember that there may be more of a build-up of material here so note the difference in thicknesses to outer areas. Older buildings may have been worked on for many centuries, so the search for original material (or what is deemed original) can require some intelligent assessment, sometimes necessitating highly specialized assistance.

During work to the Manitoba Legislature, a condition survey of the massive bronze statue atop the dome was needed. The dots were for taking a laser scan and the open panel was for an endoscope examination of the inside structure that revealed the mounting bar was rusting away.

TELLTALES

Used to measure movement in walls, the concept is that there are two fixed points where the difference between the first and subsequent measurements will indicate the distance by which a gap is widening. I have come across many telltales installed in buildings, and while it is a good idea the problem is that the original measurement always goes missing. So here is a quick method for a telltale that shows the movement without the need to keep track of records.

Take a pair of microscope slides and with an adjustable square for accuracy, mark a set of lines (towards one end) on the glass with a TCT scriber or chisel on each slide that match up perfectly with the marks on the inside when paired up; wipe a bit of black paint or ink into the scratches and stick a date tag on the end. To put them in place get some quick-setting epoxy putty, sold as plumbing seal repair, mix up two blobs and stick the ends of the telltale to the wall over the fracture, ensuring the marks are matched up. Any movement will cause the marks to go out of alignment, thus telling the tale.

Telltales in Salisbury Cathedral. One is a cement block; the hope here is that it cracks when any movement takes place. The other is composed of two pins that have to be correlated with a record somewhere.

Slides with marks scratched on, to make a budget but very effective telltale.

Telltale set in place; the putty squeezes up and sticks well. When writing on buildings, pencil is probably the best as it does not fade or wash off; if this is an exterior position protect the information somehow.

Medieval arch has been set out using unevenly sized voussoirs which give it a unique character. The journeyman stonemason would order enough stones to fill an arch of dimensions that he would give to the quarryman who, not too concerned on regularity, supplied unequal-sized blocks to make the voussoirs. The stonemason would have to work with this material and not be concerned with the irregularity, so he would apply his section template to the joint and make the moulding meet in the middle. This produced the irregular face pattern seen here.

These marks on the clerestory handrail in Coutances Cathedral are significant and historic; they are rope burns from pulling up stone and other materials during the building of the cathedral.

Analysis

First and foremost is the use of your eyes and brain. Look at the specimen and work out what is in it. The following points are a real help for recording items and specifying what is to be done:

- Nature of the wall and its method of construction
- Materials used
- Original mortar and style
- Intervening work and materials
- Sources of decay
- Structural condition
- Condition of wall materials
- Extent of mortar degradation/cause

When the nature of the problem(s) is known, then the work can be specified and should take these questions into account:

- Does it require structural work/support/ties?
- Are all moisture/rain dispersal systems in good order?
- Is removal of old work necessary? If so, how much?
- Is the new mortar suitable in performance/colour/ texture?
- Are there items which are apparently damaged but should be preserved for their historical significance?

Two examples of honest repair using ferrous metal add to the character of the stone, but will eventually cause problems. These need to be noted and commented on, possibly giving a time limit in which they should be inspected and possibly rectified. The issue is that they are part of the history of the whole, so should they be taken out, treated or ignored?

PAINT SCRAPE

For onsite examination of paint layers a scrape is made. With a sharp scalpel blade, cut into the paint at a low angle, down to the original surface, to remove a sliver about 10–15mm across and leave the layers of paint exposed as elongated sections. Lightly brush and examine with a pocket microscope; if possible take a natural light photo at high resolution on macro setting for a detailed look.

CROSS-SECTIONS

If the call is for detailed (microscopic) visual examination or micro-chemical testing of a mortar or paint sample, it is handy to mount it in a solid block of resin with the sample set so that the components (mortar) or layers (paint) are presented in cross-section.

Place the sample on edge in a plastic container (make sure that the container is resistant to the solvents in resins) with a good area projecting above the edge; if necessary support by a mini-frame of toothpicks and a blob of clay/plasticine. Useful for paint is a sloping cross-section, which allows a greater area of each coat to be presented, so if possible set a sample at 45 degrees as well.

Use a clear, slow-cure resin to fill the container around the sample, gently agitate to remove air bubbles – if available use a vacuum chamber – and leave to set. There are numerous online tutorials for making a DIY vacuum (degassing) chamber; choose one that falls within budget and is also practical for mould making, using easily obtainable items – it is even possible to harness a water hose instead of buying a dedicated pump.

Once cured, remove from the container and grind the projection off flush with the top of the resin. On a dead flat surface (sheet glass or metal) place a sheet of 250-grade wet and dry paper, dribble some water on and start polishing the section. Subsequently use finer and finer grades to get a highly polished face, always rinsing off the debris from previous grades to prevent scouring.

The finish will not be truly visible while it is wet, as the water will mask scratches, so dry off the sample and examine it with a magnifying glass or microscope during the process to assess progress.

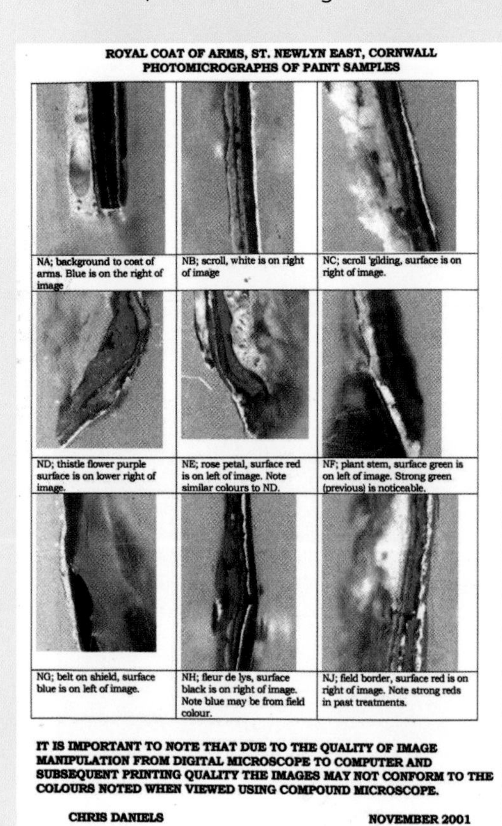

ROYAL COAT OF ARMS, ST. NEWLYN EAST, CORNWALL
PHOTOMICROGRAPHS OF PAINT SAMPLES

NA; background to coat of arms. Blue is on the right of image	NB; scroll, white is on right of image	NC; scroll 'gilding, surface is on right of image.
ND; thistle flower purple surface is on lower right of image.	NE; rose petal, surface red is on left of image. Note similar colours to ND.	NF; plant stem, surface green is on left of image. Strong green (previous) is noticeable.
NG; belt on shield, surface blue is on left of image.	NH; fleur de lys, surface black is on right of image. Note blue may be from field colour.	NJ; field border, surface red is on right of image. Note strong reds in past treatments.

IT IS IMPORTANT TO NOTE THAT DUE TO THE QUALITY OF IMAGE MANIPULATION FROM DIGITAL MICROSCOPE TO COMPUTER AND SUBSEQUENT PRINTING QUALITY THE IMAGES MAY NOT CONFORM TO THE COLOURS NOTED WHEN VIEWED USING COMPOUND MICROSCOPE.

CHRIS DANIELS NOVEMBER 2001

ROYAL COAT OF ARMS, ST. NEWLYN EAST, CORNWALL
PHOTOMICROGRAPHS OF PAINT SAMPLES

NK; crown jewel, blue surface is on the right of image. There is metal leaf visible (as yellow dots top half) this may be from the crown itself.	NL; plume, surface white is on left of image.	NM; lion body, this sample is polluted with other colours, metal leaf is visible at top centre.
NN; surface is on left of image.	NP; inner frame, black surface is on right of image.	NQ; outer frame, black surface is on right of image. Wood is the rectangular section.

IT IS IMPORTANT TO NOTE THAT DUE TO THE QUALITY OF IMAGE MANIPULATION FROM DIGITAL MICROSCOPE TO COMPUTER AND SUBSEQUENT PRINTING QUALITY THE IMAGES MAY NOT CONFORM TO THE COLOURS NOTED WHEN VIEWED USING COMPOUND MICROSCOPE.

CHRIS DANIELS NOVEMBER 2001

Paint cross-section sample sheet from a report. Set in resin, the samples show the layers of paint from each element of the artefact; the light material with the air bubbles is plaster.

Typing Up

Back in the office, get all the notes out and start putting them in order. The report format should be straightforward, contain all the relevant information and be easy to understand. Each section (chapter) needs a number and every paragraph or sub-section in it should have sequential reference numbers that start with the section number; the reasoning is that correspondence or discussion about any aspect will be easy when all holders of a copy can locate the exact point.

Report Contents

Here is a typical layout of a report; use it as required. Years ago the photographs and diagrams would be in an appendix at the back; nowadays it is possible to insert these into the document where necessary.

- *Introduction or Abstract.* This will set the scene. Name the structure and the purpose of the project/report. Introduce the people involved and give a brief outline of the project.
- *Description.* Describe the structure in detail using correct architectural and historical terms, dates, location, materials of construction, environmental situation and present use.
- *Condition.* Analyse the state of the structure, what is happening to it, the extent of decay/problems and the causes of decay.
- *Recommendations.* State what needs to be done. Prioritize needs.
- *Specifications.* Pinpoint how to do the work, and the materials and processes involved.
- *Observations.* Assessment of how the work went. Problems and results.
- *Maintenance.* State what is needed in the future to keep the structure and the work carried out in good repair.
- Glossary, Suppliers and Reference.
- Photographs and Drawings.

Once written and finished, print off copies and put them in good-quality folders with a cover title and hand out. Digital copies are also welcomed nowadays, but make sure the formatting is fixed and will work in other operating systems – saving the document as a PDF file will do this.

Get Organized

For good-quality work to be carried out on site, it is best to be on top of all the intricacies required for things to go smoothly; these all need to be in place before work starts, so at the time of survey start thinking about how all the logistics and day-to-day life on the project will mesh.

Access

Starting from the outside, check there is vehicular access for loading or suitable (and affordable) parking close by. This is an essential as many things involved in building projects are heavy or awkward; so can materials be dropped off on site? Will parking have to be paid for and/or what are the restrictions? Is the scaffolding up to the job? Can the areas to be worked be accessed without too much hassle, and will the spreading out of kit and materials cause problems? Make sure that any hoisting gear is safety checked and strong enough to cope with whatever is needed. Ensure the water and electricity supplies reach to the work areas, and that there is enough water pressure to get up high enough; check the hose fittings – make sure they are compatible with personal kit or buy adapters.

The Mess

Where is the waste disposal and who is paying for it? Is a skip needed or can the rubbish be bagged and taken off site; remember there are transport restrictions on commercial/trade waste as well as fees to be paid at tips. Work out how much protection to the property is needed in the form of sheeting and masking of windows. If there are planted areas below the work, consider the trouble of getting debris out of the vegetation and if necessary protect it or provide alternative run-offs.

Onsite materials will need to be on pallets or similar to keep them dry; for example, powder materials are best decanted into large plastic tubs with lids; do not forget to indelibly mark the contents on the side (not the lid).

Mortar production and working of stone will create copious amounts of mess unless carefully organized. Plywood sheets with battens on the edge will prevent debris getting on the floor and be easier to keep clean. Hessian or Terram matting laid beneath pallets will trap dust and fragments, while

Typical leftovers from working with mortars. Here at height they must be bagged and got to the ground, which could take some time, so costing should allow for this.

The possible problems with leaving a site like this unattended for any length of time are manifold, ranging from theft or damage, to a risk of injury to children playing on the poorly stacked stones in this open churchyard.

allowing water to drain through. Be aware where waste liquid ends up, as slurry will quickly block standard drain systems, and water containing fresh lime (when cleaning the mixing area) can irrecoverably stain any materials it comes into contact with; if possible, have a hose with a jet spray to hand to dilute and disperse any staining liquid. For any heavy washing work, rig up aprons that lead into guttering. Have a downpipe going into a water butt to allow any debris to settle out before it goes into the domestic system.

Security

Making sure that nobody climbs onto the scaffold or steals anything is always an issue, so have a routine for removing and storing first lift ladders out of reach. Tool trunks should be attached to scaffolding with locks or similar and have security locking.

Restrict public access to the site with hoarding or temporary fencing, such as Heras. There should be signage to warn people of the dangers and provide deterrent information. All regulation notices and warning signs must be displayed, as legislation requires. These should be in laminated waterproof cases and securely fixed to the work areas.

Health Matters

A first aid kit and eyewash (absolutely essential for working with lime) should be on hand at all times, with relevant contact details of medical help displayed.

Appropriate insurance should be obtained and displayed for any work, so check out what is needed and get it in place.

There will always be a need to get oneself clean, so determine how this is to be achieved, either on site using the building's facilities or installing a set-up for this; this is a preferable option to driving home in filthy footwear and clothes (although overalls are a boon). One option which I found came in handy was when I worked in a city with good sports facilities nearby: a subscription allowed a shower, sometimes a relaxing swim and a cup of coffee before setting off home.

Selection of Stone

There are many considerations when choosing a particular stone for use in replacement or repair. Obviously it would be preferable for the original type of stone to be used, but often it has to be a compromise by using stone that matches the original as closely as possible. This need not be a problem as there is a huge variety of stones available. The task is to recognize the qualities of the original and find a suitable replacement.

King's Statue, Weymouth. A superb Coadestone group on Portland stone plinths was cleaned of decaying paint using pressurized steam, repaired and repainted with lead-based oil paint in naturalistic colours. The decision to use gilding, instead of the previously cheap alternative of yellow paint, was based on evidence from examination of the layers removed.

The trademark of Coadestone on the King's Statue, Weymouth.

Usually described as an artificial stone, Coadestone is actually a ceramic product that should be classed as stoneware; its other name, Lithodipyra, should be a clue as it means 'twice-fired stone'. It incorporates grog (ground-fired clay) to reduce shrinkage. The quality of Coadestone statuary and architectural elements is outstanding. From its original formulation by founder Eleanor Coade in the eighteenth century, it has been used by major sculptors and craftsmen and has proved extremely durable and hardy. Cement castings and finishes are often mistaken for this material; conversely Coade is often mistaken for stone.

The Egyptian House in Penzance is still described as having Coade decoration, which has always been questionable (the quality of the sculptural elements could not compare to the usual detail of Coadestone). Samples were taken off and looked at; knowledge of Coadestone manufacturing methods made identifying it incredibly simple for us. Typically, higher authorities were uncertain that workers could identify the material that well, so an expensive analysis was commissioned, with the same result.

The knack for identification is that Coadestone is packed mould clay, whereas cement mixes are basically liquid; there is always going to be some air trapped in a cement mix, so if there are any round bubbles (empty spaces) in the matrix then it is definitely not ceramic. The other identifier is the construction method, as Coadestone is never thicker than about 5cm, so it will have walls of the same material to support the larger shapes. Dunting in the mass will also help identification.

Coadestone (and occasionally other makes of stoneware) is often encountered on historic buildings or gardens, needing repair. On inspection identify it correctly, as stone indents are not recommended; mortar repair using fine sand, calcined clay and hydraulic lime have been utilized on site. The other method is to model up the replacement piece in the same clay mix, but 7–10 per cent bigger to allow for shrinkage, and to fire this in a kiln before fixing. This is skilled work so visit a Coadestone workshop and have a chat. Indeed, always try to visit interesting manufactories; if they are skilled and know their work is good they will be comfortable chatting with other artisans; those who know they are hawking a shoddy article will be less inclined and more secretive.

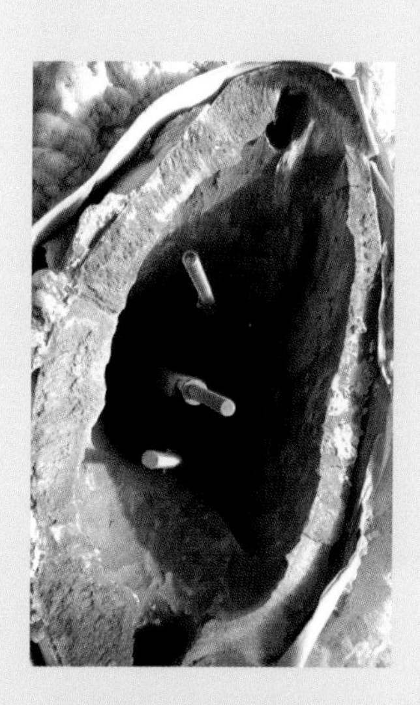

Inside the torso of the King. His head was held down by an iron rod, but this and other iron inserted when the statue was erected had rusted out; a stainless steel armature was designed and inserted. Note the thickness of the Coadestone.

The dubious quality of the Egyptian House sculpture came under professional examination to determine whether it was really Coadestone as claimed in the official records or, as discovered, some very coarse cement-based modelling. To be frank, anybody who has actually looked at Coadestone with a critical eye would realize that modelling of such crudity as found here would never have made it out of the factory door.

A personal sample collection for reference and demonstration to clients is easy to build up with a bit of legwork; all suppliers hold these samples and are keen to get their product sold, so get their contact details and ask for some blocks to be sent.

The Bibemus quarry in Provence, used since Roman times, still bears the quarryman's tooling on the exposed stone, which was worked *in situ* to get a flat surface prior to breaking it out of the ground.

Samples taken from quarry blocks are marked and ready for testing to see if the stone will stand the exposure it will be subjected to in repair work.

The majority of smaller quarries were usually opened up for local use, and most vernacular historic building was carried out using local materials, so if there was only a limited amount of the stone available, the source has probably dried up. The importing of specific stones for fancy or fashionable work was difficult and expensive, usually limited to grander buildings; on the up side, many of these stones are still being quarried. We must include the caveat that if the original quarry is still producing stone that has the same classification and name as the original it can be different in performance, texture and colour. Quarry and supplier information for stone will basically have the same disclaimer stating 'stone is a natural material and may vary in colour and quality from the samples shown'. This is important to know; you may need to visit the supplier and pick out suitable stone.

Choosing the Stone

Replacement stone must match the characteristics of the original, so it helps to know the qualities required by a stone and how these compare to what is available; stone suppliers are required to give the data of their product and this will go some way to helping. There are stone collections such as the one in the Natural History Museum in London than can help in the hunt; a good working knowledge of stone is essential to narrow down the search. The Stone Federation of Great Britain has all major UK stone suppliers as members and can often be the first and last source of information regarding the obtaining of stone; other countries will have similar organizations.

This roof shows an unusual use of slate, with large slabs butted up to each other and a slate bead fashioned to cover the joint. This would last for ever if it were not for the rusting of the nails blowing it apart.

Identification

With experience some stones can be identified by eye; others may need to be compared to samples or tracked down by knowledge of local geology. The levels of investigation can become quite complicated but most of these can be learnt and practised if there is a desire to do so; workshop science is feasible right up to producing thin section samples for microscopic examination. It is up to the artisan to follow this route and find out where suitable training can be acquired.

As an avid lover of gadgets, the chance to own useful ones is always a bonus to me. Here are a few microscopes that can help in this work; the tiny one is always to hand in my bag and great for looking at inclusions and micro-fossils; the workshop one is for more detailed investigation such as paint samples; the binocular or stereo microscope is old but serviceable (what can wear out on these?) and useful for looking at grains in three dimensions and getting depth. Note the plastic Geocard for grain sizing and shape reference.

As this is a general practitioner's manual we will concentrate on the sedimentary stones, the most common material encountered in buildings. Most people identify sedimentary building stone to be either sandstone or limestone, and sometimes it can be difficult to distinguish between the two, especially if they are firmly attached to a building, as the effects of weathering may disguise the material's original colour or texture. Two simple checks will suffice for the immediate job, though as some stones can have some characteristics of both it is best to carry out further work to pin the correct stone down. Marble and granite can be instantly recognizable by colour and inclusions, whereas slate, flint and schist tend to be local stones and easily tracked down in the area.

The first is a visual check which is related to pollution staining, so it can be done from photographs. It depends on how the stone is stained; when the upper, rain-washed surfaces and ashlar are stained, the discoloration is due to oxides in the material reacting with (acidic) rain and rusting, thus indicating sandstone. If upper surfaces are clean and ashlar dirty, with a crusty build-up in the places sheltered from rain that ranges from light toffee colour to black, then the stone has a carbonate binder and thus limestone is indicated; the (acidic) rain dissolves out the calcium when it converts by reaction

Two very similar-looking stones are tested for calcium carbonate by dropping acid onto them, with differing results; one is sandstone, the other limestone, although looking almost identical in appearance and texture.

Comparison of new stone to the effects of soluble salts on similar pieces through moisture travel.

THE MOHS SCALE OF RELATIVE HARDNESS

The Mohs scale is used to determine the 'relative' hardness of a mineral, or something composed of minerals (that is, stone). The test requires no laboratory equipment, is simple, and is often done in the field as a primary analytical method. When the material will scratch at some point, you have a number to work with.

1	Talc	6	Orthoclase
2	Gypsum	7	Quartz
3	Calcite	8	Topaz
4	Fluorite	9	Corundum
5	Apatite	10	Diamond

PRACTICAL APPLICATION OF THE MOHS SCALE

Whilst the standard kit is desirable it may often be impractical to transport or keep in a toolkit; there is a practical way to overcome this by assigning hardness value to everyday items – see table (right). If possible confirm that these items do actually have the same hardness as the minerals in the test kit.

To ascertain the hardness of a mineral (stone), start with the hardest implement from your test kit and work down until an implement scratches the test specimen; then assign the specimen number lower than the instrument that scratched the specimen.

EXAMPLE

If a knife blade scratches the specimen, but not a penny, assign it a relative hardness value between the two test instruments, in this example a relative hardness value of 4, the Mohs scale value of fluorite.

This is important and very useful as one can then look in a reference manual to discover what minerals (which visibly match the specimen) have a relative hardness value of 4. Often this rudimentary determination is sufficient to aid in field identification of minerals.

PRACTICAL FIELD TEST KIT

Test kit item number	*Item in kit*	*Relative hardness*
1	Fingernail	2.5
2	Penny	3.0
3	Steel knife blade or window glass	5.5–6
4	Hardened steel file	7
5	Emery cloth	8–9
6	Diamond	10

to water-soluble sulphate in areas where it can soak into the stone – under cornices and in carved details – and this effect spreads when unchecked.

The second method is very easy, but only verifies calcium carbonate in the binder, so one could reasonably assume the sample is limestone (though calcareous sandstones do exist). Scrape a tiny area of the stone clean, blow the dust away and put a drop of acidic liquid on the surface: brick cleaner is ideal (vinegar can be used but it is a slight reaction and denser stones may not react). If it foams the sample contains calcium carbonate – the binder of limestone.

History

Obviously building styles and use of material are affected by a country's history, so a working knowledge of architectural history is a boon and can be picked up as you go along; a quick bit of research starts any project off well and gives confidence when dealing with other participants. The UK is a prime example of changing styles and materials: the country's maritime history has meant that stones as ballast in ships have been brought in via sea ports, styles have been seen in many other places and been copied, and the eleventh-century Norman invasion led to the import of stone from Caen in Normandy for innovative major building projects, simultaneously giving us much of our modern working terminology.

Colour and Texture

These aspects of stone may be self-evident when it is newly cut; however, the final colour may well be different as it can change when dry or weathered. Inspect abandoned faces in the quarry for some indication of colour change and scrutinize buildings constructed of the same stone; look in protected areas for less weathered examples. New stone pieced in can be substantially different in colour and often in texture (it helps if some texture is added when fixing), so it will be noticeable. The effects of pollution, decomposition and past cleaning give the original stone an altered appearance – this is generally more noticeable with sandstone than others. Do not make the mistake of using a stone for repair or replacement purely because it has the same colour as the existing; over time it will invariably weather and decompose differently, with often strikingly bizarre results.

The texture of a stone when quarried and cut is due to the minerals it is composed of, pore structure and manner of deposition; this will alter over time due to the effects of weathering, loss of material from the matrix and its level of exposure. Even texture and regularly sized grains are not to be taken as an indicator of physical strength or durability.

Dartmoor pump house, converted to a dwelling. All the stone for the arches was found buried in the ground: it just took a bit of dry assembly and detective work to sort them out for their respective arches.

Tower in southern France, built of cretaceous limestone that is obviously local to the district.

The correct stone for this replacement coping has been selected but sadly the quality is suspect due to the clay venting across the top; it will fail quite soon.

Durability

The durability of a stone will usually be affected by many factors. As these will be different for every usage (and often the stone itself will not have a uniform composition or structure), durability tests should not be considered as having universal advocacy.

The standard test for durability is by soaking the stone in a solution of sodium sulphate and letting it dry; the subsequent crystallization of the salt will start to break down the stone. Over a number of cycles the result is expressed as a percentage of the stone compared to an industrial standard, and it is only that cube that counts! As a rule of thumb, coarsely textured stone tends to be quite durable, unless it has poorly cemented beds and inclusions of clay that weather out quickly and lower its strength. Denser limestones let less moisture in and can be resistant to the elements, but are used less due to the difficulties of hand working.

Depth of Bed

Sedimentary stone size for masonry blocks is limited by available bed height, which is the depth of a homogenous layer of stone as a result of a period of deposition, or where different materials or geological factors produce a different stone layer between. The other factor is its natural joints which occur in stone before quarrying, where breaks or lines

New coping stone laid on edge (note the direction of the shells in the stone) to prevent any delamination of the beds, which is possible without the pressure and restraint of stones pressing down; this is the best option for this situation.

Calculating the height of bed to be got out of a sandstone block destined for export to Canada. Note the redder inclusions midway down; these are lumps of clay and must be discarded to make the stone good for masoning – the size you see is not necessarily the size you get.

of cleavage perpendicular to the bedding plane result from geological pressure or movement; it is important to know the dimensions available when specifying or designing a piece.

Availability and Cost

It is obvious that there must be sufficient quantity of a particular stone of the correct size that can be delivered within the necessary time. The period from order to delivery is the lead-in time and is crucial when setting up a project; that this can be shortened by sweet-talking, inside help or cost adjustment – all business practices that may be considered. For rarer stone, detective work may be needed to source second-hand blocks in reclamation yards or unused, over-ordered supply secreted away in a quarry or masonry yard, that can be cut to size; it always pays to have a 'nose about' whenever the chance comes up. There are also cases where a stone is available but the methods used in quarrying, such as blasting, mean that the majority is unusable; the disproportionate amount of wastage also causes a large increase in the projected cost.

Stone is a fairly cheap material; after all it is everywhere! The real cost of stone comes in moving and processing, and this soon mounts up; so when pricing for work remember cost of stone sawn six sides (SSS), delivery to workshop or site or both, masoning time, how to move it around site and fix it; and do not forget the taxes either.

Tightly bedded steps in a Roman building have small nubs on the risers; possibly these were used for lowering them into place.

Testing

Tests on the porosity of stone and its effect on durability are useful but will not always be pertinent, as it is the internal construction of a stone that is significant in its reaction to decay factors.

Various agencies have tested all dimension stones available for all relevant performance standards, so comparison of these results and perhaps selective testing of the preferred stone may be of use. When accepting stone deliveries, test for soundness by striking (a strong tap with a metal bar or chisel) the block. If it sounds dull compared to the others it may include a vent or other fault in the stone; in some cases it is possible to break the stone along the fault and pin it back together, bearing in mind the masoning and structural implications.

A good-looking block of Hamstone which only shows its fault when cut; here Saul points out the almost invisible vent that reduces the size of stone available from this piece.

CASE STUDY: IT'S NOT BRAIN SURGERY

Statues have singular issues, being exposed on all sides with little bulk to absorb and deflect the elements. They are also prone to neglect and inept repair from well-meaning but unskilled local tradesmen employed by their owners. Another issue is that the members of the public watch the statue out of the corner of their eye as it fades and dirties up, accepting the shabby state as giving it historic kudos, so when restored to a decent level of cleanliness there can be hullabaloo about the different aesthetic it now holds. When writing the report this issue should be covered and the patron made aware that there may be a significant change to their 'patinated' artefact.

The head had at some point been knocked off and shattered, then stuck back together with crude cement repairs that had turned a hideous colour.

Carrying out the primary inspection to assess condition and requirements of the Sir Henry Edwards Statue, Weymouth.

Once the project was started, the head was removed and the torso cleaned down with pressurized steam and mild detergent.

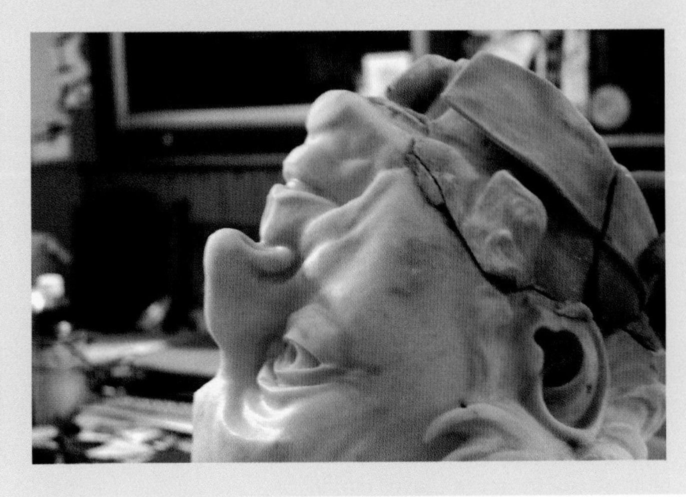

Back in the workshop all the cement was removed and the marble cleaned, then the fragments were pinned back together in roughly the same way as the finial repair described elsewhere in the book.

Open joints and missing areas were made up with white epoxy resin putty, polished in to blend in; the colour, though noticeable in close-up, would not be seen once it was up high. This is a lesson worth understanding and remembering: you may have to convince the owner while standing on a scaffold with non-practitioners making derogatory comments about being able to spot the work (because they are an arm's-length away), implying a lack of skill. Indicate a noticeable mark, then take them to the ground and ask them to point it out.

The bronze dowel can stay, and lead softening strips are put in place to space and support the head while the mortar sets.

Refixed, the head now has a crown of stainless needles set in a mouldable mesh and epoxy putty skullcap; this deters seagulls and keeps him cleaner for longer.

On the King's Statue in Weymouth, as my first attempt at bird deterrence measures, I had made a lead skull cap set with stainless pins; here the cap is shaped to the head.

Unfortunately it was viewed up close first and this predisposed an official to think it would be an eyesore, even at 12 metres above the ground; in truth it was hardly visible unless deliberately looked for and was extremely efficient.

The revised model applied to the King's Statue. While efficient, it does not protect the paint from the elements as well as the original; still, he who pays the piper calls the tune, so it had to be done (actually the taxpayer was paying, but that is another argument).

CHAPTER THREE

THE WORKPLACE

The scaffolding at Castle Drogo in Devon is going to be one of the largest such projects in Europe.

OPPOSITE PAGE:

Many years ago the keep in Corfe Castle, Dorset needed to be inspected cheaply without scaffold (no access for cherry-pickers) so it was decided to abseil down. The problem was that there was no way to get up to anchor a rope, so a bow and arrow (with twine attached) was produced and many hilarious attempts were made to fire a line over the top. After some time it was conceded that this had failed miserably so the local coastguards were contacted and obliged by firing their grappling line over the wall, saving the day and earning much refreshment at the local pub afterwards.

Most of the work covered by this book will be on site as it is mostly related to buildings and they tend to be hard to move! Talking about workshops may therefore seem superfluous, but this chapter needs to be interpreted according to where you are based for the job. The main aim is to have a dry, secure place with the space and facilities for storing and caring for equipment; if it is possible to carry out work on the premises as well, so much the better. However, we first need to consider the most important piece of equipment – yourself.

PPE is all correct, although there may be an issue with the method of mixing mortar carried out here.

Looking after Yourself

Everyone has a duty of care for their own personal welfare and, hopefully, a sensible desire to work in a safe manner; this can be achieved if practice and the working environment are conducive to this. It is the contractor's responsibility (in law) to ensure that anybody working on a project is kept safe from harm; this can be achieved by providing a suitable environment, detailed instruction in the use of the workshop and, if necessary, training in how to use equipment.

Train yourself to be as methodical as possible on site and in the workshop. Lay out the tools needed for a task and ensure all are fit for purpose. Have a clear workspace or bench and lay any softening/protection down before getting materials organized.

Manual Lifting

Always know the weight of anything heavy, and do not be afraid to ask for help in lifting and moving things; it is possible as a fit youngster to manhandle seriously large stones but the toll is paid in later years and working life may be reduced. Learn how to lift correctly to prevent back injury or hernia.

Trapped fingers are common when moving stone, but they should not be; always have a batten under a stone when placing it, or quarter it on the edge of the banker. And wear gloves! Sack trucks, trolleys (with big wheels for uneven surfaces) and rollers are all there to make life easier; keep them in good condition and always secure unstable items when on rough surfaces.

Dangers of Dust

Be aware of the nature of the stone. It may be necessary to get some protection from harmful mineral content such as sandstone silica and fine powder materials. Personal protection for this ranges from paper dust masks to all-encompassing pressure-fed helmets where the forced air provides a clean atmosphere to work and breathe in. The simplest method to prevent inhalation is by using a dust mask but working for hours or days wearing a protective mask is not a pleasant situation (or practical, as they become less effective each time they are removed and replaced in the course of the day). On the other hand the prospect of contracting some debilitating and possibly fatal lung disease is very unpleasant, so some research is vital on the most appropriate methods available. In the best scenario, an industrial-quality dust extraction system would be installed. The costs and maintenance involved will make this only viable for those that can afford it, but portable extraction units can be hired. Remember that cost should be secondary to quality in health terms, so factor this in when pricing.

Bankers and extraction in Salisbury's workshop, ready to begin work.

Well-constructed mobile scaffold platform ready to work off but, as it does not give quite the right access, a dodgily angled ladder is being used.

Working outside is obviously a good idea, though it can be anti-social to cover property and vehicles in dust, and it is very unpleasant when the weather is bad.

The Eyes Have It

Dust that can irritate eyes or flying fragments that cause pain and possibly injury will literally abound, so suitable eye protection is absolutely essential, the emphasis being on *suitable* in the absolute best possible sense, as eyes are not replaceable commodities. Decent goggles or high-quality protective glasses are available everywhere now, so always have a pair or two to hand. Keep them in a protective case or bag when not in use, and be careful when cleaning them; rinse them off with clean water or by compressed air and try not to rub the (abrasive) dust into them. Lenses made of glass rather than plastic remain clear longer, as stonedust will not ruin the surface so readily.

Shading

Working on stone outside, during daylight hours, is a good choice, especially if it is sunny or bright. If the stone is pale then it is noticeable that a mild form of 'snow blindness' can occur, giving rise to headaches and eyestrain; in this situation a pair of tinted glasses should be worn – remember though that they still need to be of safety standard.

Feet

Sensible footwear is a very important item and should always be worn. Anything that falls down could land on your foot at some point, and there are a lot of things waiting to prove the attraction of weighty things for the down direction – from relatively lightweight chisels to the blocks of stone. Good quality comes into its own here as well – these will spend more time on your feet than any other shoes, so comfort is a big consideration as well as insulation and durability. For those rare sunny days on the scaffold it is possible to buy sandals with protective toecaps, and the coolness of working in them is lovely, though they tend to fill up with dust!

One from our workshop archives; nattily aproned workers, overseen by my predecessor Herbert Read, replacing a statue at Exeter Cathedral.

A busy site can soon become cluttered, but here there is order in the way tools and materials are kept to hand; risks and mistakes increase without due attention.

Clothes Maketh the Mason

Get a decent pair of overalls, preferably several, because it is sensible to clean them regularly, thus protecting everyday clothes, the interior of vehicles and your washing machine. They should be loose enough to allow warm layers underneath in the winter and to keep you cool in the summer; white is a good colour, as it will not show limestain so readily. Kneepads are essential for all work on the ground; get gel-filled versions with straps that do not constrict movement or circulation. A workshop apron is a good thing for slipping on to do small tasks or carrying dirty materials.

Organizing the Workshop

Keep It Clean

Be aware that the best-laid plans for an immaculate workspace with everything sorted and ready for action will not succeed without self-discipline. The reality is that after a day/week on site, you will arrive in the workshop tired, dirty and wanting to relax, so all the equipment and materials will be unloaded and literally dumped off as quickly as possible. Do not despair – we all lapse; just get into a regime that suits your personality and keep it under control.

Stonemasonry workshop in use, managing to look orderly with everything to hand.

Personally I go through the workshop once a month, or during a day rained off, sorting out the rubbish; reuse aggregate sacks or buy heavy-duty rubble sacks for this – do not try bin liners as they will split! I then have the space to check power tools and untangle cables, clean hand tools and oil/sharpen/repair them as necessary.

Two essentials for a good workshop/shed are weatherproofing and a decent floor. Follow this up by providing shelves and hangers for everything, and try to use them!

Kit Care

Chisels can be stored in boxes or tins, with a piece of rust-reducing oily rag to rest them on. Larger tins can have a layer of lightly oil-soaked sponge on the bottom which protects blade edges.

Power tools come in cases, so dust them off and store the equipment in these – there is usually enough room to keep bits, blades and discs in there as well. Buy spare mechanical tools – the spanners, screwdrivers that are used to fix these – and secure them in each case; this will be a huge help when an essential drill, for example, needs a quick repair and you can lay your hand on the right tools immediately.

Wipe handsaws over with a lightly oiled rag and hang them up; there is a hole somewhere on the tool for this.

Old lifting chains will last for generations if cleaned and oiled regularly.

Remove all dried mortar/plaster (especially as plaster is acidic and promotes rusting) from trowels with scrapers and wire brush. Rub them smooth with wire wool and oil the blades.

Before working on stone, always wipe any oil off the tool; stains will get into the surface and are very difficult to remove.

Buckets should be rinsed and brushed out as soon as they are not needed; if this is not done they become unusable very quickly and then the dried and hardened material needs to be beaten out. If this happens, use a mallet or piece of wood and pound the outside to detach the residue but be aware that hitting cheap plastic buckets (especially in cold weather) will fracture them; use good-quality rubber buckets, trugs and mixing containers that can take abuse.

Useful fixings, drill bits and odds and ends used on jobs are easily lost, or thrown away during cleaning if they are lying around loose in toolbags. Purchase stackable plastic containers or storage trays or recycle coffee tins and screw-top jars and keep them topped up with consumables.

Light and Space

Headroom and space must be taken as it comes when securing a workshop, but if opportunity allows use a space with substantial height. This allows a lot more work to be carried on before it gets too dusty and will allow for the installation of useful lifting gear. A high space is also more pleasant to work in.

The cost of scaffolding can be prohibitive for some jobs, so roped access work is becoming a trend, shown here at Carcassonne; training for this is expensive but opens up a whole new aspect to the services offered and may be worth considering.

Underfoot

The floor should be smooth, level and hard, preferably concrete or with well-made solid board/timber planking sturdy enough for anything needing to be moved about on wheels or rollers. Working and moving on uneven surfaces for long periods is uncomfortable, tiring and dangerous.

It never ceases to surprise me how much debris and waste is generated for any type of stonemasonry. The accumulation will take regular cleaning, usually needing a shovel (and hopefully an industrial vacuum cleaner), so if the quality of the floor is low this process will get worse as you continue, so the floor needs to be of the best standard you can achieve.

HAPPY FEET

Once the floor is up to the task of surviving the rigours of a busy workshop, you will find that standing for hours on this hard, unyielding and possibly cold surface will create discomfort that affects the rate of work; try to stand on a cushioned surface such as carpet or a rubber mat. Now a recognized issue, specifically designed mats are available for all-day working in such circumstances; standing on a square of carpet will make all the difference to your comfort and keep your feet warmer for longer in the workshop and on site as well.

Banker

A fixed banker (of sturdy nature as it has to take a lot of weight and punishment!) can be made from cementing or pinning concrete blocks together to form a solid mass of an appropriate height. Use steel bar for dowelling the blocks and pieces of carpet laid in the joints to absorb vibration. Consider that there will be much pounding and vibration so ensure that whichever construction method you use can take the punishment. Portable bankers can be bought, but it is not that hard to get one knocked up out of treated timber by a carpenter, or indeed make one yourself at home. It needs to be moved about and lifted onto scaffolding so it should not be too heavy, just sturdy.

A useful banker would be about level with the waist in height, have a working area of about 60 × 60cm and have a shelf underneath for tools and a bracing rail at the base which will give a footrest whilst working. When working on a scaffold a ratchet strap can be looped over this and under the planks below to improve stability. Alternatively use a 'hop-up' – a small sturdy platform purpose-made to stand on and work from; these, made in a variety of materials and styles, should always be robust and large enough to be safe. Whatever the work surface of the banker it will need some form of softening to protect the stone edges and prevent the block sliding around; carpet squares from a furnishing shop are the most economical and practical method of covering the work surface, although they do collect dust.

Portable bankers and workbench, strong yet light enough to carry around, made to order by a local carpenter.

Heavy Lifting Equipment

Aluminium gantries can substantially speed up jobs and earn their keep; there is a variety of lifting equipment available so

Temporary banker made from a tree trunk; the rope stops the stone sliding about while being worked.

HOLDING SMALL THINGS

If the stone is small it may be difficult to keep it in one place whilst being worked on; this can be resolved by filling a washing-up bowl with sand and resting the stone in it. This is useful for carrying out small carvings that need to be shifted regularly or do not have a flat surface to rest on. Sandbags will perform all these functions as well, although you cannot use the contents for mortar afterwards; inner tubes filled with sand are another good alternative.

Hoisting stones into position on site in Ottawa shows the need for heavy duty kit and always having helpers around. (Photo: RJW-Gem Campbell Stonemasons Inc.)

do some research. A mechanic's engine hoist or a fork-lift trolley can do the work of many men in lifting and placing stones; compare the cost (spread over years) to the cost of hiring help, and as an asset it will reduce tax bills. If the investment is too much, do not be afraid to hire plant, as this allows you to get kit that is up to date and of a higher standard than is often bought for personal use. Use correct straps in good condition for lifting stone, with working clips. Good-quality rope that can be knotted and undone easily for hauling or tying up should be kept neatly stored; jam the loose end between the handle and body of a bucket, wind the rope around it and secure the end for keeping it handy but not in the way.

The cost of dedicated lifting apparatus is repaid by the speed of moving stone around, and less time is spent with the chiropractor.

A Place for Everything

The workspace should be as uncluttered as possible; this standard may be hard to maintain when the work is in full swing, but the workspace should be tidied at least once a day. Toolrolls and clever boxes are great for keeping everything neat in the workshop or car, but tend to be emptied and just shuffled round the workspace; sort out which tools are to be used and put them in buckets, as these are easily carried around and tools can be dropped into them and stored at hand when not in use.

The workspace can quickly become very cluttered, so good housekeeping is essential to ensure efficient working.

Not **the way to leave a site; anyone could wander in and damage the work or, worse, themselves.**

If waste is being generated, line a plastic tub with a rubble sack with the top open so that larger bits of debris be easily dropped in as they are produced. Have some cable ties to secure the tops when they are full, as this makes moving them around so much easier, cleaner and quicker. Brush up every day and bag the dust, so that the next day you have a clean start, thus preventing the mess building up to time-consuming proportions.

Electric junctions for extension cables should be protected from rain and water spray. Cut holes in a suitable tub with slots to slide the cables down; place the connection inside, put the top on and hang the whole thing up to prevent it being kicked over.

Have battens or softening (non-absorbent so it does not become a sodden mass when it rains) to store stone on, and provide a cover if necessary. Try to do all mixing at the designated area, and only take enough material/mortar to last the work period; keep the dry materials in sealed tubs and ensure these are marked on the side (tops can get mixed up) as materials similar to each other can get confused. Always have a sprayer and two buckets of clean water to hand when doing any work with mortar.

The Toolkit

The possession of a decent toolkit should be every artisan's goal, and there is real enjoyment in trying to accomplish this. With the rise of DIY, craft courses and hobbyist activities, there is a myriad of tool suppliers available these days; the issue is, which tools to get? The simple answer is: those you need and of the best quality – without them you cannot do your job. Owning and using really good-quality tools gives great pleasure.

The tool selection here, and mentioned in other chapters as they are appropriate, may cover old ground and some may be so obvious as to hardly need mentioning, but just as I like owning, organizing and using tools, the chance to think

Here the site has the correct security fencing waiting to be erected, which once in place allows insurance to be valid.

A stonemasonry pitching hammer has an elongated head with a small strike face, giving lots of force to the hit.

about them has given me more ideas about what may help to make working life satisfying.

New or Old

When buying tools, unless a catalogue can be trusted or you are buying on recommendation, it is necessary and more rewarding to handle them, checking the quality of the workmanship, material and fit. Always go for high quality and respected makes. A good maxim is 'Only buy tools once'; this is a lovely sentiment but even if they can wear out get the best you can. Metal parts should be neatly finished with no burrs, scratches or dents; machined edges need to be exact; parts that fit together must do so without slop; and the fixings (screws, bolts and rivets) must be up to the job (adjustable setsquares are notorious for making errors if poorly made). Stainless steel is the choice for rules and straight edges, forged metal for trowels (definitely not stamped or welded here), and well-forged steel for chisels. Wooden handles should be securely fixed with strong ferrules, oiled and varnished.

Buying second-hand tools is a good way to build a kit on a budget; certain good-quality tools will outlast a working life and can reduce the expense. Check the condition; a bit of surface rust is not a worry but breaks in blades or chipped edges should be shunned if there is not enough meat to get back to a working level. All wooden handles can be replaced easily; a better handle is often all that is needed as a good upgrade for hammers and mallets.

Classic range of stone-working tools, bought second-hand; all will last a few generations more.

Setting Out and Measuring

A good set of drawing instruments made of durable material (steel) is preferred to plastic, which will become fractured, scratched and unreadable very quickly. Tape measures will come and go, but a couple of sizes are essential marked in both metric and imperial, with a tape lock. They must be robust as they are much-abused items of kit. Laser measuring devices are useful for large hard-to-access areas and noting for quantity surveying. There are also levels and setting-out laser equipment, and all are useful. Steer clear of really cheap ones and buy from a reputable manufacturer; the added expense will be an incentive to look after them properly.

Metal straight edges and squares are essential for all work; try engineers' stores or sales rather than expensive specialists; and if buying second-hand check they are true. Place the straight edge on a flat clean surface and, with a sharp pencil, make a line along its length, then rotate it 180 degrees and draw over the line – these should match up exactly; repeat this process, starting in the same way, then turning it over and drawing over the line to check. Place a square against the straight edge of a board and draw a line along the inside of the blade (the outer edge is not used for marking out) on the board, then turn it over and place the corner at the start of the first line and draw another along the blade; these should match up exactly.

Sinking squares and sliding bevels are used for checking the depths of fillets and the accuracy of chamfers between them, as well as for transferring setting-out information to the stone itself.

An old but good square will outlast the modern sliding square.

Engineer's sinking square or depth gauge can cope with chamfers as well.

Tungsten scribers can be made from old chainsaw files, which come with a handle and are usually thrown away once they do not sharpen the saw anymore. Shop-bought ones are similar in design to pens and can be clipped in pockets. Very hard pencils are a must, H9 or H10; do not get carpenter's pencils as they will not last for more than two lines before needing sharpening again. It is possible to get hard lead fills for propelling pencils, though they tend to snap easily and are fiddly.

Curves

Compasses and dividers should be sturdy enough to use on stone and metal; beam compasses or trammel heads that fit on battens will be needed for big curves. Plastic template sheets and French curves are used for setting out moulds and developed sections, while profile gauges and lesbian curves or similar can be used to transfer detail.

Cutting

Good pairs of tinsnips, preferably at least one pair with curved blades, are handy if using traditional zinc for templets; nowadays though it tends to be plastic film so some good scissors and disposable blade knives will do, as these will also manage lead sheet.

Hard point saws will cope with soft stone (as well as wood and plastic), and if possible a mason's saw with TCT teeth is a good addition for harder stones. Hacksaws and bolt croppers are useful for cutting dowels without power, though metal cutting blades for disc cutters are inexpensive and can handle a lot in a short time – always remember to check the type of blade in the cutter if it is used for this. Cutting discs are not

Scribers and stonemasonry scribe.

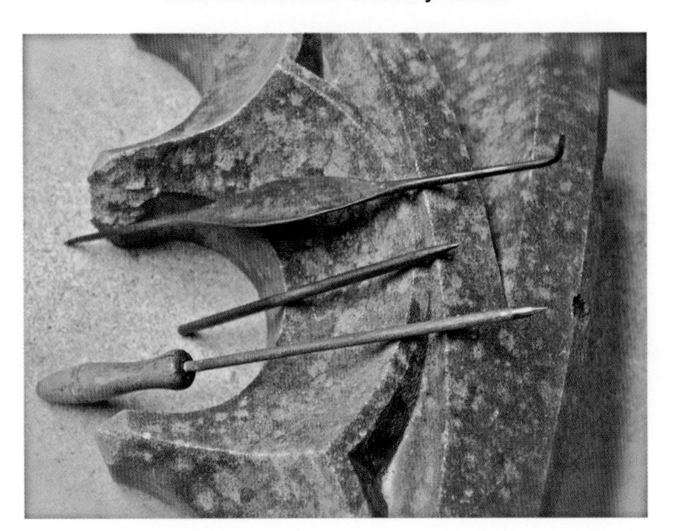

Snips and tough scissors, old and new, work well when of decent quality.

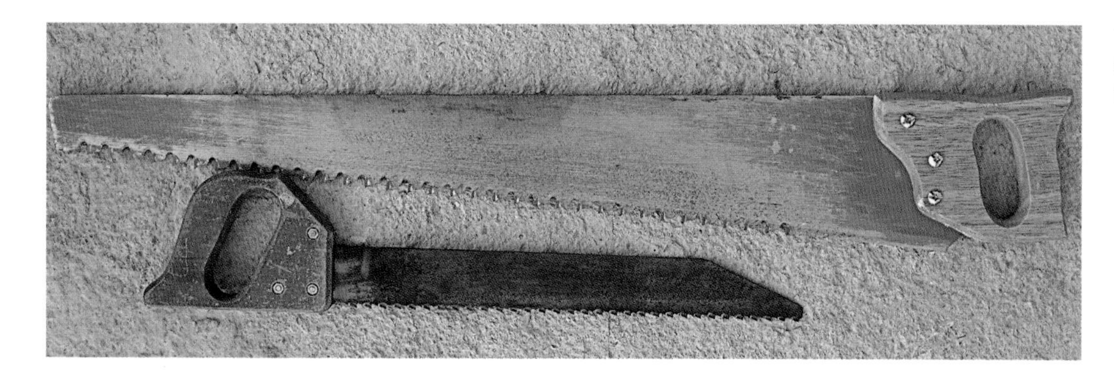

TCT stone saws; the smaller is for cutting fillets.

designed for grinding or cleaning metal, so have a special blade for this to clean the burrs from newly cut bar.

Stone Kit

For general work on stone a surprisingly simple toolkit is all that is necessary. Nylon mallets are the ones to have, being easier to use and maintain and lasting longer than traditional wooden ones; they come in a variety of sizes, so try to get one suited to your needs. Hammers are used by many in place of mallets and are cheaper to buy, even the correct type for pitching and punching stone; the range available is wide and they have many uses depending on the head and weight.

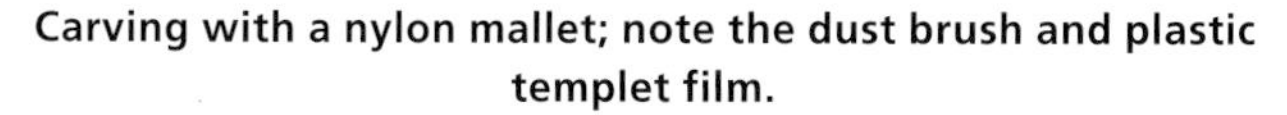

Carving with a nylon mallet; note the dust brush and plastic templet film.

Stonemasonry chisels are not the same as those available in most shops, so find a supplier and get at least one of the following from each group shown here.

Punch or point; it is possible to use one from a standard SDS (heavy duty hammer) drill kit.

Patent claw, with spare blades. Get two sizes of blade, 5cm and 2.5cm; some of the latter can be ground to give a curved cut for channelling out.

Pitcher (handset). Here a TCT version is being used for granite.

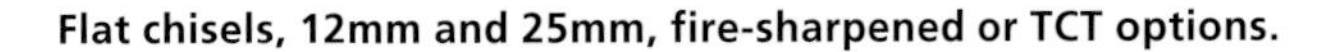

Flat chisels, 12mm and 25mm, fire-sharpened or TCT options.

Boaster about 5cm wide; TCT is best.

Tungsten carbide tipped (TCT) chisels are the most durable and will keep an edge longer than forged steel (fire-sharpened) blades. Do not worry about these having a round top (mallet-headed) as with a nylon mallet the end of the chisel will not damage the head as much as when using a wooden one. For softer stones, specially made wooden-handled chisels can be used; these are more easily sourced on the continent than in the UK.

Bullnoses 2.5cm and 5cm; fire-sharpened or TCT options.

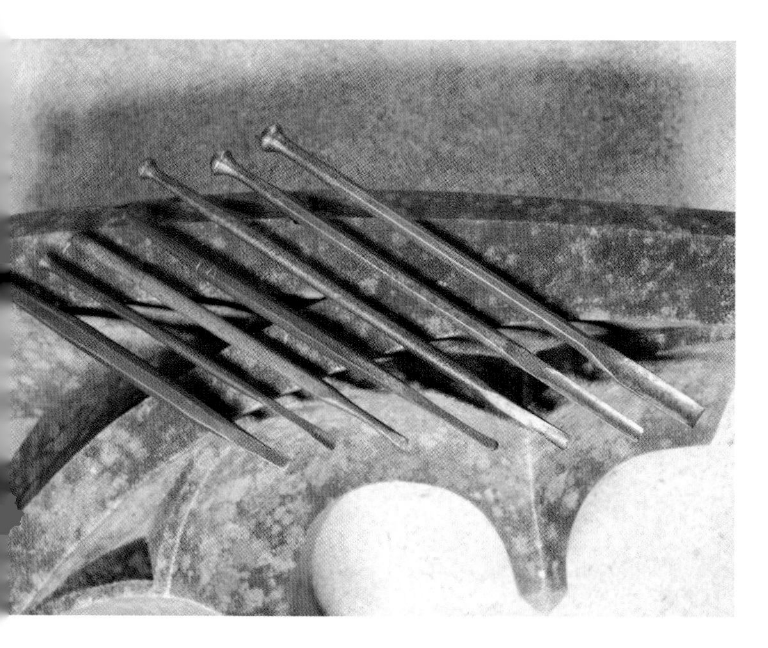

Other items which may come in useful include carving tools, gouges and lettercutting chisels.

Bathstone (or French) chisels used for softer stones.

Lengths of various types of saw blade can be accumulated for dragging stone or raking out joints, but be aware of where the point is going when being used. It is helpful to put a gaffer tape softening around one end to create a handle for the joint work.

For instant site repairs to spalled stone or fractures, have a couple of tins of polyester adhesive (sold as stone glue) handy, stocking liquid and thixatropic versions to cover all situations; remember, though, that it is essential to add pins if the repair is outside.

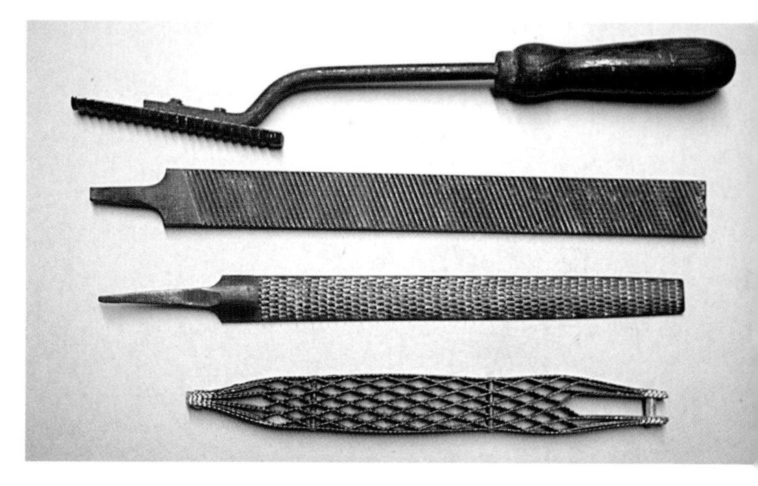

Files for stone; the top one is for drips, the others are from other trades.

Air Tools

Pneumatic hammers are widely used nowadays in stonemasonry and can be a boon when they are used correctly, though overuse can cause a serious, debilitating medical condition known as 'hand–arm vibration syndrome' or commonly 'white finger'. Obviously a compressor will be needed for these tools, but once purchased this can be used for many other tasks on site and in the workshop. Pneumatic disc cutters and grinders are all available as alternatives to electric ones. Spray cleaning guns are good for fine misting of mortars and backs of joints, and clearing dust from holes and joints is easy with an air blow gun.

Not modern art, this is traditional artisan scaffold on the shore of the Red Sea, ready to start construction.

It pays to have a sturdy bench grinder with wheels suited to steel and tungsten for shaping tools, also for chisels prior to sharpening; the final honing should be with a whetstone or diamond sharpening block. Mount a cup holder next to the grinder for quenching and cooling hot metal but remember the safety implications of having water near electrical equipment.

My personal range of trowels covers all uses in mixing, fixing and shaping. Some may prefer other types, so consider what people say and have a go with different versions.

Fixing and Site Tools

The fixing of stone into a building, especially one that is already constructed, needs to be conducted with the right tools. Spirit levels and straight edges are obvious for getting work to line up; look after these so that they remain accurate and do not bend them. Chalk line and plumb bob will both need to be used; to prevent a wasted hour of untangling always store string lines securely wrapped round a bobbin and secured by a rubber band; five minutes doing this properly will free you to earn more money.

Trowels

These are available in a range of shapes and sizes so have a selection of these: small leaf or spatula trowels used by *stuccotori* (ornamental plasterers) are for fine pointing; a pointing trowel is a small diamond-bladed trowel with the sharp end rounded off for coarser pointing and manipulating mortar; gauging trowels and bucket trowels are used for knocking up and shovelling about; pointing trowels with long thin blades are preferred by some for fine work. A trowel must have the blade and tang (the bit that goes into the handle) made in one piece, not welded together as this will create a weak section and will not last.

Using a large GRP hawk to build up mortar over a bespoke ventilation system.

Hawks

These are made in three materials: wood, aluminium or fibreglass (GRP); the last is the one to go for as they are light, durable and have the best friction surface for picking up mortar. Big hawks are good for lading and using large amounts of mortar but can be a bit cumbersome to handle; the smaller ones are indispensable for all manner of work.

Lifting a slab with a pinch bar. Note the softening to the stone, and the knee pads making life at ground level bearable.

Old-school slate chopper. With practice the most delicate shaping can be carried out with this; it is excellent for using slate in a tile-style repair.

Odds and Ends

A portable banker or solid workmate (there are some wondrous site benches with all manner of gizmos) could have a range of modifications allowing tools to be stored underneath, with possibly a mount for a vice and clips to screw down through to anchor it. A handful of hangers like butcher's hooks or bent lengths of bar can be latched onto scaffold tubes: these will keep drills and disc cutters off the floor, hang buckets out of kicking way and always let ear defenders and goggles come easily to hand.

Shavehooks, putty knives, tungsten scrapers and high-quality stripping knives, as well as a good range of paint-brushes, cleaning fluids and paint kettles, are not just for decorators; they all belong in this toolbox. Nail or pinch bars of about 30cm long will help to position stones and allow lifting strap access and many other jobs; remember they need to have softening on the lifting end to protect the fragile stone – layers of gaffer tape will do at a pinch.

A general carpentry kit is essential. This will start with a claw hammer, a dedicated saw (the one used for stone may be getting a bit blunt by now) and a battery drill/screwdriver (remember to have a spare battery and take the charger along), backed up by a big box of assorted screws, bits, washers, rawlplugs, spare fuses, staples and every other thing that runs out or is needed to keep work flowing. At the last count I have six of these and will be getting more. Other items that come in handy for bashing stone or surfacing are drywall and scutch hammers.

If you come across an old slate chopper, buy it and practise with it. Tile repairs or packing can all be done with neatly cut bits of slate; and any roofing tasks that need trimmed slates will show the advantage of these in adept hands, compared to the modern scissor version.

So many loose wrapping jobs, protective sheeting and covers are needed on site, workshop and home. The securing of these can be immediately sorted with a stapler; buy a robust heavy-duty version and have a good supply of long staples to get through multiple layers; use stainless steel staples if corrosion is an issue.

It is probably not any surprise to learn that carrying a decent mechanical/universal toolkit has sorted out many a problem on site, especially with power tools. This should include fuses, spare plugs (they can squash easily), insulating tape, jubilee clips, cable ties, circuit tester, twine and wire amongst other things. I know a planning officer whose car is loaded to the gunnels with every bit of kit going, and she takes great pride in being able to sort out site problems with the kit of the burly workforce.

Guns

A caulking gun (or two) is needed for applying the wide range of fixing and adhesive resins, silicones and other mastics that are used so often on site. Always have spare nozzles for materials in these tubes; get really heavy-duty ones as they break or seize up at the most inconvenient moments, and keep them

An engine hoist makes a perfect crane for heavier stones such as this sundial plinth.

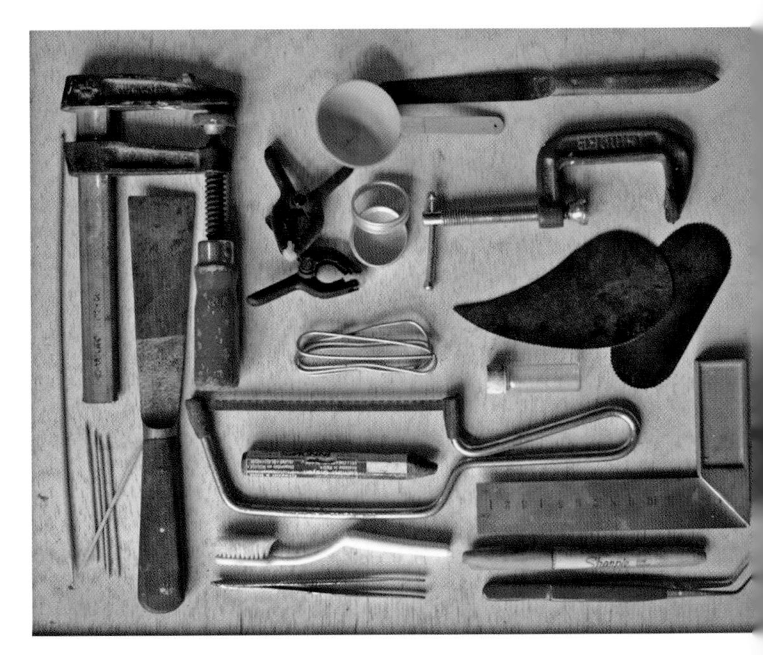

Various useful bits of kit include paint and palette knives, cockscomb and kidney scraper for finishing and clay work; clamps (including mini-grips), very useful for attaching sheeting together or in place; tweezers, marker pens and crayons for marking out work or stone; containers for interesting stuff; toothpicks that can have cotton wool twirled around the end for that conservation cotton-bud effect; a junior hacksaw; and old toothbrushes.

A selection of eroding tools that I have found useful for finishing and blending in; these can be found at boot sales or good tool shops (if buying second-hand, check for blunt cutting edges in the centre). Curved rasps and flat metal files without handles can be held flat against the surface. Below are various rifflers and a tungsten carbide rubbing block.

clean and maintained. Resin injection kits will have a special application gun for the particular cartridges used; these are pricey so should be kept clean and lubricated regularly. Buy a well-established make and hopefully the refills will be easy to procure, as running out of an obscure type can halt a job or entail buying yet another applicator. Always overstock with the nozzles of any squirted stuff.

Wedges

Wedges of various sizes will be needed in many situations and should be made as folding pairs; basically take a rectangle of wood and cut it diagonally to produce two identical wedges. When placed long side together and moved in they will lift evenly. Made from hardwood, sanded smooth and possibly waxed, they will slide easily and are waterproof to boot. Make them as needed when you come across a suitable piece of wood; always knock up a few sets as they get lost and damaged in the day-to-day business of running a

Modelling tools can be found at art shops or specialist plasterwork suppliers. The picks are bought as dentistry tools – ask your dentist for old ones. The strange dumbbells are for making locating dimples on clay; they can also cut out perfect circles in paper or film or even metal – hold the paper over a hole in the metal, place the ball in to this, pressing to locate, then strike with a hammer (a boon for gasket making!).

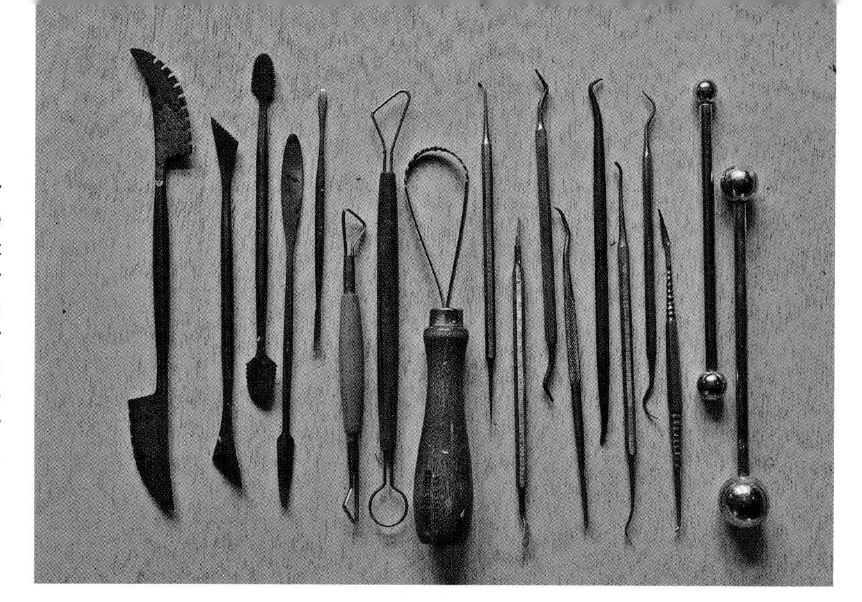

Brushes never go out of fashion and, like shoes, one can never have too many! Wire brushes, though not recommended in the text, come in handy for many tasks including cleaning tools and preparing dowels; rotary ones can be for a drill or, even more aggressive, a disc cutter. Pipe or bottle brushes come in all sizes so have a range to suit the dowels being used. The hoof brush has a pick that can be cut off if it gets in the way or you can grind it to shape for a specific job.

Masking tape, insulating tape and PTFE (plumber's) tape are not as handy as gaffer, but will be just what you need on some occasions. Cable ties will hold up something leaving your hands free. The Velcro© sticky tabs are great for mounting protection that needs to be lifted for inspection. Superglue is just that; be aware once opened it will be useless one day so check it occasionally. The ribbon materials are: rubber strip for softening, gap insulation strip for sealing large joints while grouting or protecting smooth surfaces from slabs, as is the sticky-backed compressible foam rubber strip. The white block is surfboard foam (ask for free samples from the suppliers) and can be cut and shaped easily – you will often find a use for this, and especially for the two-part epoxy putty in the blue bag. The lesbian curve not only takes shapes but can make an adjustable support for many tasks such as positioning a hot air gun to blow in the right direction.

Oddments needed on site are always diverse; here are a few essentials. Barrier cream will protect and prevent lime burns and also make cleaning easier. Always have a bottle of rust treatment, it will be used. Keep some eyewash even nearer than in the medical station; dust bits flying around can be irritating so if you have itchy eyes wash them straight away. The tissues and mirror are helpful here; being able to look behind stones, etc., is always a boon. The blower, though not powerful, will help to shift dust. The scraper is a universal tool with a tungsten blade; it can get in tight spaces, smooth out lumps and do many other jobs. The wire snips are essential and will be vital to getting an electrical tool going one day. The white mesh is cotton coated with a thermal melt plastic and can be used to take shapes or make an armature at a pinch.

Disposable gloves come in boxes of 100; buy from farm supplies outlets – they are tougher than the discount tool store variety. Nitrile are the toughest but are attacked by solvents, so layer them with latex if needed. Clingfilm can do no wrong, so always have a roll handy. Spare glasses wrapped safely may never be needed – but when they are, it will be crucial. Syringes and pipettes start getting into conservator territory. The serious injector is great for pumping small amounts of grout or consolidant; it is normally for animal husbandry so can be bought in farm supplies outlets again (they have many useful materials that are much cheaper than in specialist shops, so always have a look round the shelves). The plasticine and Rilem tube (porosimeter) are explained further on. The Tinytag is a data monitor and can record thousands of temperature and humidity readings (which much easier than using a hydrometer and thermometer), all of which can be downloaded and produced as readable information for inclusion in reports or to check the environment in a building. The puffer brush is a camera lens cleaner, but helps with the fiddly stuff.

If paint samples or stone examination are needed, here is the starter kit. The microscope is a professional one and not cheap, but you only need to buy it once in a lifetime, and it impresses when you are able to get this sort of detail when describing material. Use the Geocard for describing grains. The glass plate has very fine wet and dry polishing paper stuck to it for a flat surface used to prepare samples. The little blocks are paint cross-section samples and the slides are thin section stone with the thickness measured using the vernier gauge. The lenses in the foam are polarizing to identify minerals and the plastic tray can be used as moulds to set paint samples in resin.

Paired folding wedges made from some imported hardwood found in a skip.

job, and mark and keep them in pairs. Buy bumper bags of plastic spacers in assorted thicknesses, as used by builders and window fitters; they will be indispensable for most fixing work.

A ladder is useful but for real access without the scaffold, a cherry-picker is needed; nowadays certified training is needed before one of these can be hired and used.

In the Bag

Armed with the above you are now properly equipped, so now it makes sense to look after everything by getting a means to carry all the hand tools about and keep them secure when not in use: a solid toolbox/chest (to sit on, rest stone on or use as a hop-up) is essential. It should be as tough as possible, as the life of the stonemason's toolbox is a hard one. As it will be dragged about on many surfaces and be hoisted onto scaffold, solidly attached lifting handles are more useful than wheels, and a lockable lid is better than leaving multitudes of drawers and trays ready to spill out if knocked over. Obviously work involves moving about a lot so a good-quality toolbag with lockable zip, handy pouches and hardwearing base will be a practical alternative; this is where expense should never be spared.

All metal tools should be cleaned and dry when put away. Wiping over with a lightly oil-soaked rag is sensible practice, but remember to wipe them clean before using again as oil will stain stone. Store chisels and small tools in toolrolls when not in use to keep edges keen; the best choice for this is canvas as dust can be removed easily and tools will not get lost or damaged – the bottom of the toolbox can be a rough place.

Straight edges and levels should be stored out of harm's way by being hung up in the workshop. When taking them to the site a section of plastic gutter pipe with a cap makes carrying them less hassle than when loose, and keeps them in one place.

Keep It

As you should be able to identify your own possessions without resorting to micro-examination or argument, get them marked. Permanent markers only work on surfaces that are not handled frequently or areas that cannot be wiped clean with solvent. Use paint on items such as toolboxes where the removal will be difficult, otherwise hand tools need to be marked by impressing or engraving; before starting this strangely gratifying task choose symbols that can be readily identified and applied to all the tools in your box with equipment you possess – so keep it simple. A box of letter punches is cheap and initials can be stamped on appropriate spaces – or use a drill or disc cutter to cut in a design – once again keep it simple! A neat way for wooden handles or plastic is to select a piece of metal tube, with a metal object, nut or screw, tacked in; grind this back so the object and the tube are flush at the end; heat this up with a blowtorch and brand the tool by pressing it in. It is wise to insure tools against loss or theft; for proving ownership it is handy to have all your equipment listed (with receipts kept) and photographed, with close-ups of unusual details.

CASE STUDY: FINIAL REPAIR

This solid stone decoration was knocked off the gate pillar by a bump from a careless driver. As it rocked, the bottom third levered against the dowel set through the socle and into the cap, causing the lower third to split into three regular pieces.

Holes and dowels ready to locate the three sections. Note the countersunk holes.

The finial returned to its rightful place, repair almost invisible.

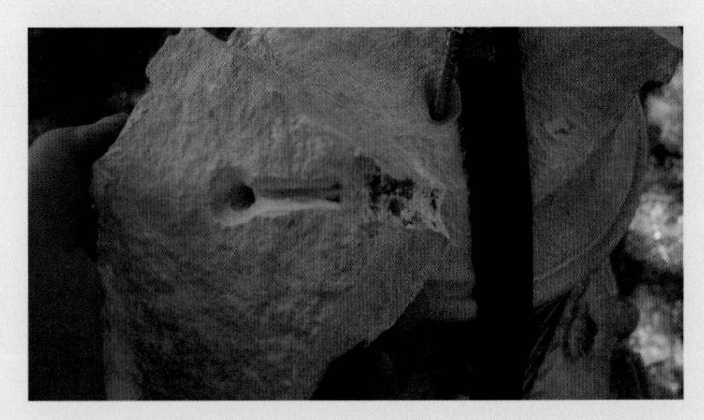

A section drilled with the same countersink and a channel carved from the centre where the main dowel sits to the repair holes; this is to allow resin to flow into the holes after it has been dry positioned. The sections are held in position by a ratchet strap with plenty of softening. Any open fracture joint was pointed up with a very fine mix and allowed to cure and then resin poured down the central hole to a depth that left enough room for the socle dowel.

Mini cramps to position across the fractures have holes drilled and channels to sink them in; the channels are made by bringing the drill out of the hole and laying it down so it cuts out the stone with the flutes.

The finial is protected by a polythene sheet and masking tape, then clay fencing is put around the cramps and the central hole plugged before resin is poured in, into which the cramps are dropped.

Resin has cured and the finial is ready to be refixed in position.

The socle dowel was bronze so it could stay. Here all is ready for a dollop of mortar and then the finial will be lowered onto the lead sheet collar which acts as softening and a spacer for the pointing up.

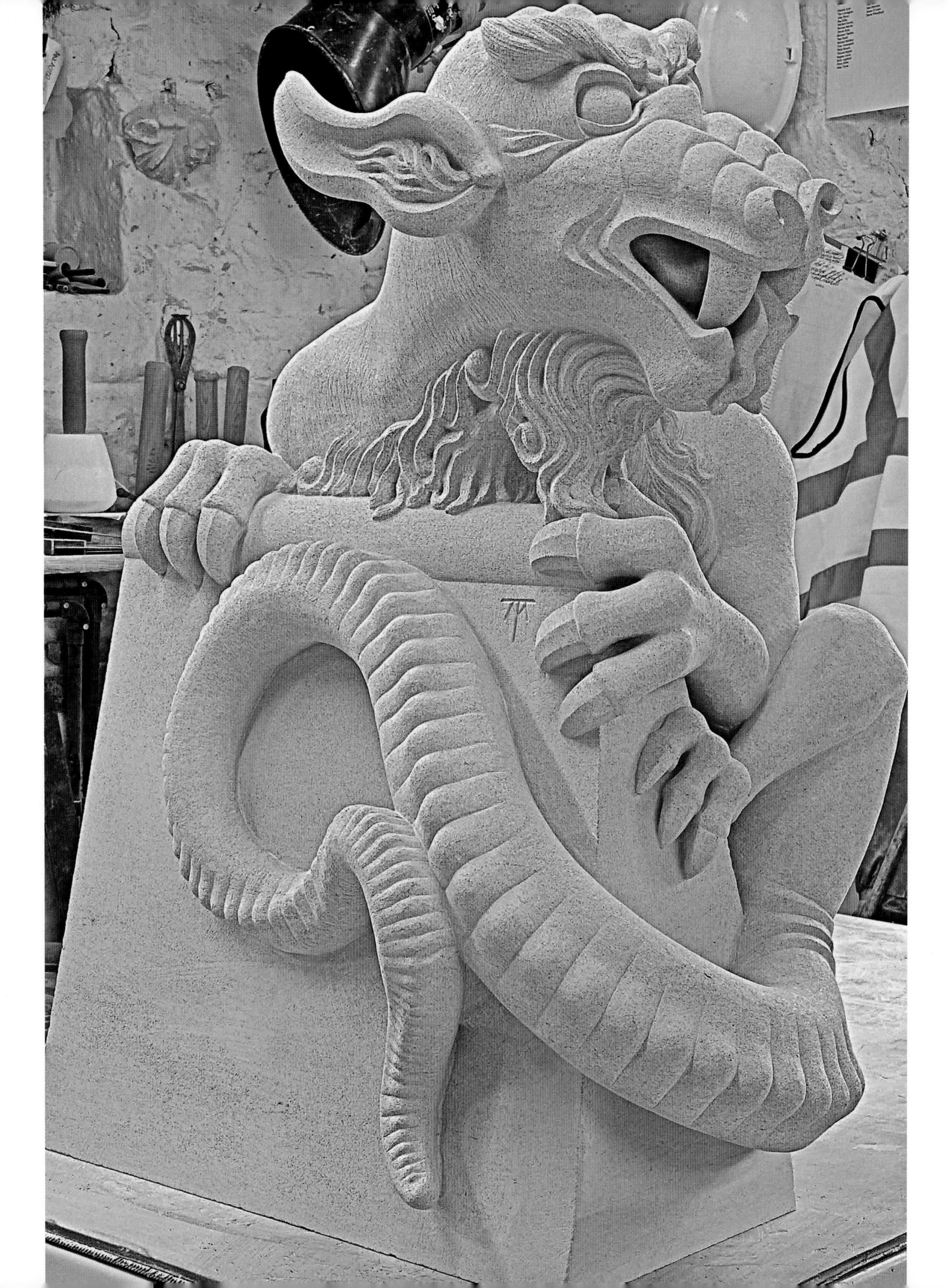

CHAPTER FOUR

TOOL SKILLS

Restoration work covers a variety of skills and techniques – here, modelling. The statue's thumb had been broken off and lost so a replacement was modelled *in situ* using epoxy putty, with a stainless pin to support and secure it. Everybody has played with clay or Plasticine – here it has been put to use.

It is necessary to become familiar with the use and care of the tools mentioned in this book. This is the first step in the craft of stonemasonry, and while it is not necessary to undergo a formal training regime, the fundamentals of tool skills must be acquired for the work set out here. Get a block of stone, preferably squared off, and have a go at working your will into it; remember all the PPE and look after yourself at all times. Have fun, yet be very critical of the quality of the work. I like to think that what is produced under my auspices, be it teaching or actual pieces, is never less than the best I can do; and so should you.

OPPOSITE PAGE:
The rich tradition of tool skills used to create our built heritage still exists all over the world; these skills just need to be nurtured and passed on to future generations. Work such as this stunning dragon carving for Lincoln Cathedral exemplifies the quality of today's artisans.

The Mallet and Chisel

Set the stone to be worked securely on the bench or banker, wedge it with sandbags, or make supports using lengths of inner tube filled with sand and tied off at the ends, so that the face to be worked on is uppermost. Standing with legs slightly apart, relax the shoulders so that the chisel side is angled to be close to the stone; bear the chisel and mallet just off the centre of the torso, keeping both level with the bottom of the ribcage. This stance allows a strike across the body, not away from it.

Boasting the surface of a block is standard practice for stonemasonry students and shows the correct position to take when working a stone, although obviously there will be times when an awkward bit requires adaptation of this.

NATURAL GEOMETRY

Throughout repair work angles, levels and verticals are used to get things perfect; to be spatially aware and capable of gauging such things as horizontal lines, to equal distances accurately by sight and to tell whether a line is straight with a glance are all skills that will grow the more you do, but it is best to always check with the correct kit. Sadly, the consequence of this is that whenever an edge is not straight, vertical, horizontal or parallel to where it should be, a niggle occurs, and more intense observations must be made to ascertain why this is.

The use of squares and straight edges is second nature in this game; after a while it will often be only to confirm what your eye and hand have told you.

Grip

Do not grip the chisel tightly; hold it in loosely clenched fingers of a flattened fist (your left if right-handed), with the cutting edge protruding from the outer edge, thumb resting on the last knuckle of the first finger. Alternatively, you may find it easier to control the chisel by resting the pad of the thumb near the head of the chisel. Without causing stress, aim to keep the chisel hand off the surface of the material, as this allows greater flexibility when cutting stone at awkward angles.

The chisel grip should be relaxed yet always able to control the chisel's orientation; if held too tightly aches will occur and the work will suffer.

The Starting Line

Determine where the first cut needs to be made and at what angle; generally this will be at 45 degrees to the bed or edge of the block. At the corner place the edge of the tool and position the line of strike so that the chisel will move in along the draft, forming cuts of the correct angle. The cutting edge

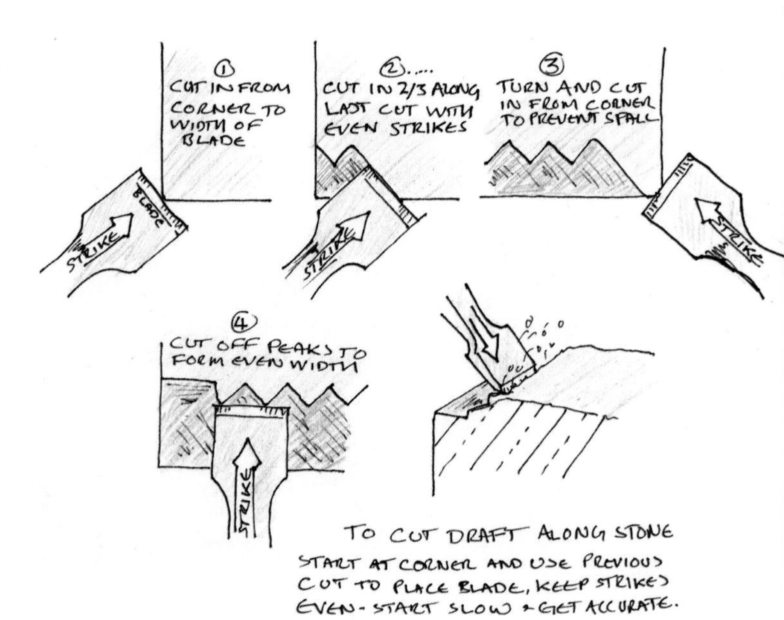

The stages in cutting a draft in stone.

Creating a diagonally tooled surface needs even working with all the lines parallel; use a 45 degree bevel to draw pencil lines on the surface as a guide.

Lining up the strike is time-consuming and a bit awkward at first, but soon the action becomes unconscious and smooth, creating regular surfaces of any shape.

should be parallel with the (desired) finished surface. The angle of the shaft to the face of the stone will be variable depending on the hardness of the stone, the width and depth of cut, and on the strength of the strike. Angled too low, the chisel will slide uselessly over, whereas too extreme an angle will result in the chisel locking in and not removing any stone. Place the mallet against the head of the chisel; slowly draw it away rotating around the elbow, then strike!

The cut is all in that first strike. Do not double tap, as this will be a waste of energy and create two variations on the cut

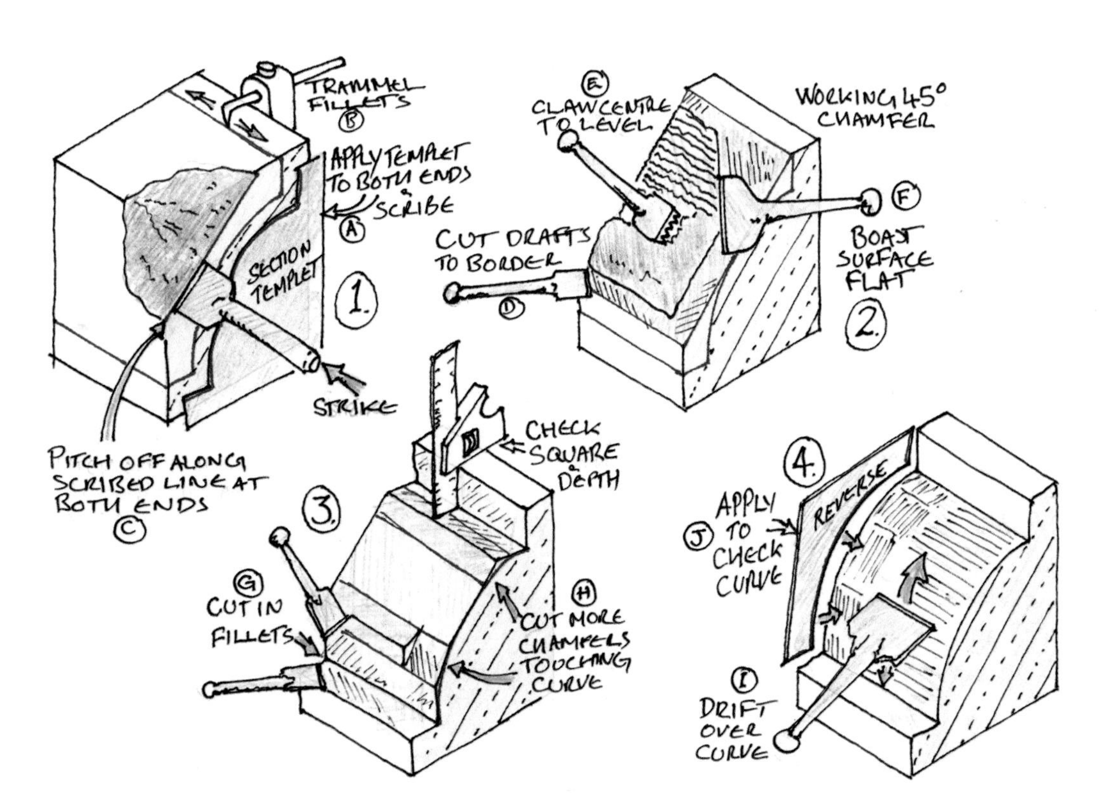

The stages in creating an ovolo moulding. Ensure the stone is square; make templet and reverse out of stiff plastic film or go traditional with zinc.

ARRISES AND ANGLES

To get a sharp arris and provide a guide for level drafts, mark a heavily scribed line along the edge and with the blade of a 12mm (hard stone) or 25mm (soft stone) chisel, set in this groove, sharply striking (technically, pitching) into the stone. Repeated along the scribe this will give a lovely even edge to work to and present as the arris. Soft or wet stone dust will cling to the blade, absorbing force and reducing the sharpness of the pitch, so wipe it after each strike.

Pitching can be developed into a fine art, with skilled artisans able to control the line of pitch to a fine degree. Scribe a line and place the pitcher edge in the groove, with the sloping edge facing away; bring up to almost normal (a right angle) to the face and strike sharply and cleanly with the mason's hammer. Practise this to know how stones will react.

– they need to be all the same. Repeat and practise until every strike cuts sharply, to the same depth and angle. Notice how the bite of the chisel is affected by all the variables coming together; the skill here is recognizing the requirements after the first cut and adjusting appropriately. Placing the chisel halfway into the line in the stone formed by the first cut, repeat this along the draft to be worked, leaning into the work without becoming overbalanced. Limit over-extension by moving around the stone rather than staying in one place, always keeping an eye to the area to be cut.

Cutting regular lines at an angle can be tricky, so whichever angle the tooling is to be, set that on a sliding bevel or use the 45 degree on a square and pencil hatch across the face of the stone.

An old but good boasting chisel has a slight angle to the blade that makes it handle better when working across a stone; the finish is shown at the corner.

Droving a Tooled Finish

The effective way is to start in the corner and, working across the drafts, start to progress over the surface hatching the face with even cuts. The task here is not to take the surface down but to put a series of cuts just below the surface.

A quick cheat to get a quasi-droved finish; with a very coarse grinding pad that scratches the stone, hold a movable edge (here the 45 degree on a sliding square) and make short strokes across the surface, moving the edge continuously to get random tool-like marks.

Draw sets of lines across the surface at whatever angle is needed with a sharp pencil, to provide a guide for the chisel. Place the boaster on the first line, hold the shaft firmly and give a strike; the trick is to bite, not cut. With each strike move the tool across the face in short regular steps, always parallel to the marked lines. Random toolmarks are generated by moving the chisel to a new location each time, while moving sideways in the same cut makes continuous lines.

Tooled surfaces do not get much better than this; its simplicity and control shows mastery of the chisel and an eye for design.

Coarse or Axe Tooling

An axe will not give such regular spacing as droved work and the blade tends to tip into the surface at one end. If an axe is used, take a two-handed stance with the leading hand gripping just below the head and the other near the end of the shaft. The blade should be between 60 and 80 degrees to the surface and held comfortably about 15cm from the stone. Lock elbows tight in to the body and drop the axe-head, with both arms pivoting equally, onto the stone using enough force to break the surface. Adjust until the required cut is achieved, then chop across the surface (drawn lines will help with aiming here).

A wide chisel can give a good axe-like finish, but it takes practice. Draw the lines on the stone and start at the furthest point to work backwards across the surface. Hold the chisel in a limp but not loose grip just away from the surface and strike sharply, drawing the mallet back quickly. Randomly move the chisel about the surface, striking here and there but always aiming to be parallel to the lines, until the surface has enough marks to satisfy.

Axing is worth a try. This tool is a tallion (its name derived from the French for 'to cut out') and was probably the major type of stonemasonry tool used up to the Middle Ages.

Using Other Tools

Trowels

The standard pointing trowel is tool enough for most pointing and mortar applications; get a good-quality one with forged shank and blade. Stick it in a vice and with a grinder round off the point to about the diameter of a penny, then clean up the burred edge with file and abrasive paper or a diamond pad.

Make up some stiff mortar and load the hawk; chop and mash this with the trowel until it becomes nice and putty-like. Next begin shaping flat rectangular patties that taper in thickness towards the edge of the hawk; with practice this whole process will take a couple of minutes for mortar that is ready to be applied. With the bottom of the trowel facing away cut into the patty parallel to the edge and deftly push the

A wedge of mortar, adhered to a spatula or plasterer's small tool, of a correct consistency for pointing; this can be applied in any orientation.

A small stiff-leaf detail, to use in casting Coadestone repair piece, modelled in clay with spatulas. If there is the chance, practise making classic details in clay with tools and not fingers; one day you could get to do some decorative plasterwork and this will put you in good stead. If your results are good enough, take a mould and cast them to keep as reference material in the workshop; they also make lovely house ornaments and unique gifts.

trowel away and up, giving it the slightest twist to support the mortar with the cutting edge under; there should be a sliver of mortar projecting from the bottom of the blade, which in a continuous movement is inserted into the joint or placed where needed.

Small Tools

These are smaller variations of trowels; some are termed leaf, plasterer's small tools or spatulas and are used the same way as the larger versions but require a bit more dexterity. They can be used for shaping mortar and plaster into quite intricate designs for mortar repair to carving; to practise this without having to make up mortar get some clay and with the various blades produce some three-dimensional models – such as a stiff-leaf crocket or similar.

Drags

These come in two distinctive types; the first have handles, with a row of toothed blades (*chemins de fer* or French stone drags) for softer stones, or a solid pad of metal with abrasive surfaces of tungsten grit or similar (the profile shaped for either flat or curved cut) to handle harder surfaces. Always have a good balanced stance, holding the handle with the leading hand; the other hand is placed on the front to guide and keep the nose down. Push these across the stone in even strokes keeping pressure on to give bite. The coarsest are excellent for removing stone quickly. Too much bite and these can judder out of line, so smooth quickish runs are more effective than slow deep cuts. To help when just cutting part of a surface down, for fillets or lower mouldings, clamp a straight batten, long level or box section to the work for the (smooth) edge of the drag to track along.

Bathstone drags and cockscombs are either steel-toothed or have TCT teeth (for harder stones) and are used to give a fine finish and should be used only near the final surface line or to knock off high points. These are held almost at 90 degrees to the surface, pulled across the stone in a series of smooth sweeps that gently take down high spots. Do not dig in and, if required, leave a finely furrowed surface that adds good texture to dressed masonry.

Chemins de fer (French stone drag; the name means 'railway line' and probably alludes to the noise it makes) with a strip of replacement blade; these are easy to refurbish and straightforward to use, though the noise is pretty oppressive.

Traditional plumb-bob is a vertical string line that relies on gravity to set up verticals; it is still used today and is handy for long stretches of stone or quoins as it provides greater reach than box levels.

Cockscombs are shaped soft stone drags and can be of any radius. Some are like French curves, having a series of sizes around the edge created by elliptical rather than circular arcs. If it is too big for a concave moulding, try placing it at an angle to the line of mould and sweep it simultaneously downwards and diagonally, flexing the wrists.

String and Chalk Lines

For fixing stones in line with others, or building new courses, string lines are pulled taut across the face, usually along the arris on the top bed (the bottom should line up with those below). They come with arrow-shaped spikes to jam into joints. The best way is to secure the spool around the corner of a face, then pull out the end and lodge that in around the far corner of the face, or tie it off on something solid or around a stone. Then place the tight line in the plane needed by putting a slip of wood against the mitre of the return and adjusting until exact. The line needs to be very taut as moving stones in and out will bang against it, so when a stone has just been placed, check the position of the line first and then bring the stone to it, using wedges and packers as appropriate.

Chalk lines also need to be anchored securely by the small

Taking off high points in a concave moulding with a Bathstone drag parallel to the run; cockscombs will be used along the run.

Without modern-day kit like laser levels and machined straight edges, the only way that structures like this could be built straight and true was with string lines and plumb-bobs.

tab, either hooked on an arris or placed over a nail in a joint. Feed the line out carefully from the spool case, keeping it taut without using fingers on the line or the chalk will rub off. Once the full extent is set, lock the spool and pull even tighter, getting the string up close to the stone. Then, if marking a long run, try to tie it off; get as close to the centre as possible and pull the string out at right angles to the surface to be marked and let it go smartly; do not attempt to twang it again as this will give a blurred line and cause confusion. If it is wrong brush the chalk mark off with a soft paintbrush, or blow it clear with the airline and do it again.

Saw Cuts

Carpentry saws are good for cutting softer stones and, depending on the set of the teeth, can give very accurate lines. To get a cut at a right angle to a surface, clamp a smooth batten to the face, and if possible one to the back, and use this as a guide while the saw is slowly pushed through. TCT masonry saws cut a much wider line and because the teeth can project sideways from the blade, a guide may need to be shifted.

Lay the guide on the line and gently work the saw back and forth cutting a line across the face, not across the corner like cutting wood, until a trench is made that will hold the blade, then continue down through; if rapid cuts are made the stone may spall out where the blade exits. Stone can have hard inclusions that catch and cause the saw to judder or jump; be ready for this and worry away at them carefully. Remember always to cut on the outside of the line.

Cutting out or making an edge for working out a mould with a hand saw. To start off scribe a deep line across the stone with a Bathstone drag held against a straight edge; use this to guide the first cuts. Work slowly and do not push too hard or the blade may flex and spall the arris.

Cutting Dowels and Bar

It is always best to use a vice to hold bar while cutting, but this is not always possible, so for hacksaw cutting clamp the bar onto a bench, preferably jammed against an immovable object. Do not cut on the handrail of scaffolding: this is against the law, can leave sharp burrs on tubes and invariably the needed piece will drop to the ground when severed. Bolt cutters are useful (although they are heavy to lug around), and when used are quite dramatic. Be aware that the severed section will shoot off and can be lost easily by falling off the lift – always embarrassing and possibly dangerous.

Disc cutters are the quickest way, but care needs to be taken. A portable vice would be best, or one clamped to a bench, but most are cut without, so if all safety warnings are to be ignored, here is the hypothetical best way. Have a board with a batten attached securely along one edge and another hinged (by a screw) so that the bar can be placed against the batten and the other pushed up to contact near the cut. Put your (safety) boot on this to jam the bar in. Hold the cutter in two hands and block your elbow against the inside of your leg (crouching with the other leg placed rearwards to balance) to create a pivot, and cut carefully through the bar. Do not cut directly onto the scaffold board, or have the cut go through the gap between scaffold boards; the blade can twist and jam, sparks will fly onto the lift below and cause mayhem.

A scale or full-size (clay usually) model to show a proposed carving or used as a guide to get depths and positions is known as a maquette. By doing all the 3D design work in an easily worked material it is possible to become more adventurous and faster at carving; it is also indispensable for showing the piece to the client and getting any design changes done before chisel hits stone, thus saving anguish and money. This is a quarter-size maquette for the replacement eagle on the Egyptian House, Penzance. It was modelled on an old photograph and the shadows here replicate the style, so the client could see that it was similar to the original.

Modelling the full-scale eagle, which went on to have a mould taken and cast in Roman cement; note the joint at the base of the neck – the piece was made in sections to facilitate mould making and fixing.

Coadestone garland on a pediment in Poole, Dorset, was fractured badly around the expansion of the rusting ferrous mounting pins; the extent was discovered after pressurized steam removal of the paint.

Summary

Purchasing this book means you as reader probably already have tool skills or manual dexterity; so the main tool advice is to apply common sense. Do not use an unfamiliar tool for the final cut; get some practice or ask questions; and always, *always* read the instructions, even if you think you know what you are doing.

Look after tools and appliances; it does not have to be every day or even every week, but get into the habit of sorting everything out into rolls, boxes and cases long before they become damaged or lost. The right tool for the job is a good maxim, but be prepared to improvise (it is called technological development in industry) if needs must; just work it out first and look at the risks – are they avoidable, will it destroy the work, break the tool or cause injury? – and figure a way to overcome them.

Safe Working Practice with Tools

Practicality in the Modern World

Unless a complete atavist, power tools (portable appliances) will become part of your arsenal and, depending on how they are used or abused, they can be extremely useful or downright dangerous; with power comes responsibility. Getting the benefits from the wide range of powered help lies entirely in the hands of the user – literally – so make sure of safety and be in control of the situation at all times.

The broken pieces were matched up and remaining cement cleaned off, then they were tacked together with fast-setting polyester resin. Once this was strong and solid, slots were cut across the breaks.

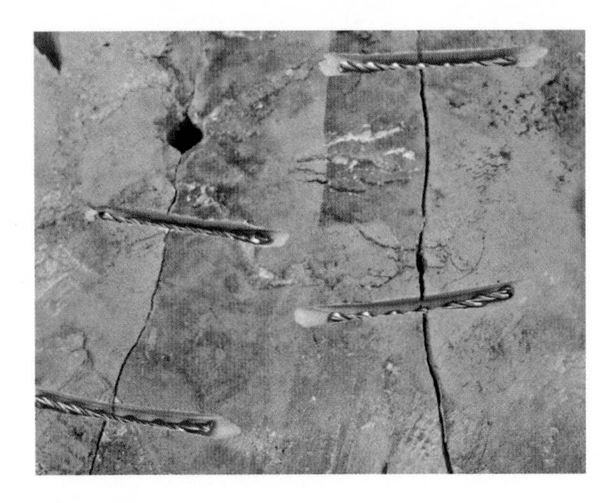

Stainless steel bar was cut and bent to fit in the segmental slots.

The dowels were set in a waterproof acrylic masonry resin, which locked and reinforced the pieces.

Refixing the sections onto new stainless dowels. When all was in place, resin was injected into the holes; when sticking together an object that has broken into many pieces, make sure they all line up correctly before applying the resin. Use support to hold the pieces into place, as any deviation will be magnified until pieces stop going together.

There is a line between intelligent improvisation and dangerous stupidity on a construction site. This situation, where an electric plug is being modified on the edge of the grinder, blithely ignores this line; if they survive this, will the ineptly modified plug be safe to use?

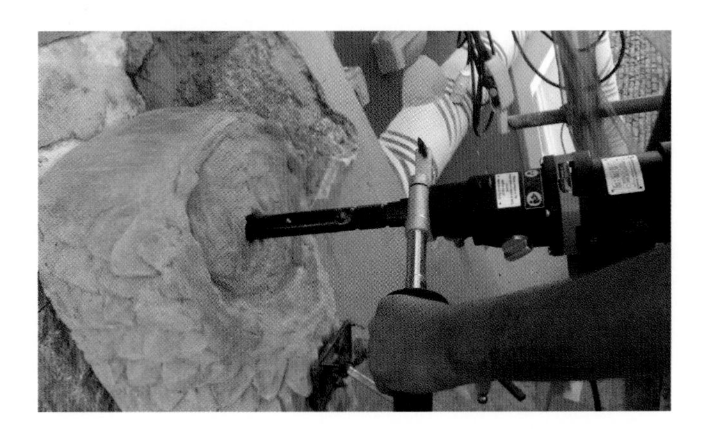

If it has got a handle to help control it, put it on and use it; the moment when torque overcomes strength is not pleasant.

Training

It is essential to use each tool properly, and in some cases training may be needed; if the tool is unfamiliar learn how to use it! This may be as simple as reading the manual or going on a training course (to change cutting discs, etc., may require a certificate of competence). Always read all the information that comes in the box, be aware of any extra on the tool itself and follow the instructions. Every tool has rules that

A really useful bit of kit is the paddle mixer, but one that requires brute strength. Make sure that you can handle the power because a mistake here hurts.

apply to it. In the case of power tools, many of these are the same for each tool every time. Learn these by heart and you will always be off to a safe start. Always read, understand and follow the instruction manual before attempting to use any power tool in any way; also read the nameplate information and follow the warning labels on the tool itself.

PPE

Get the correct kit and wear it whenever it is needed:

- Safety glasses with side protection or goggles in good condition with clear lenses.
- Dust masks for the obvious times.
- Ear protection is essential for any power tool work.
- Overalls and safety boots (with laces tied).
- Gloves are sometimes worn to reduce vibration trauma, but can hamper control and feel.
- Do not have long hair, loose clothes such as scarves, or anything else that could snag. Ensure anything that could entangle or catch is tied back, covered or removed.
- First aid kit and eyewash should be up to date, appropriate to risks, ready and accessible.
- If there is no one else around, have a phone to hand in case help is needed.

Using tools correctly can improve work rates, make money and prevent work-related injuries, all proved by the humble block and tackle here, used in conjunction with sensible rollers made from scaffold tube, allowing one man to heft a big ledger slab into place on his own.

Workspace

Set the piece up securely at the correct height to work on, remove trip hazards on the floor and have a safe place to put the tool down. Check there are no flammable/combustible materials that could be affected by hot debris. Dusty work is best done outdoors, but if it has to be inside put dust sheets over anything that should not be covered in dust. Good light is necessary. Regularly brush up and get rid of any large chunks that land on the floor.

Power Tool Checklist

- Check condition of cable for exposed wire, tears or serious kinks. Power cables have a mind of their own and always become tangled up, so get them straight and knot-free. To make life easier and improve the lifespan of cables, change them for coiled heavy-duty cable.
- Where the cable enters the tool is subjected to much twisting and often the inner wires will break – so if all seems okay and it still will not start, check this area.
- Plugs should be sound; hard rubber is best – plastic can crack easily. Are all clamps and screws present? Is the fuse the correct amp rating?
- Inspect the tool housing for cracks, missing or loose fasteners and general condition; keep it dust free by blowing it over with the airline or a fine brush after use and never drop it onto the floor.
- Discs should be completely circular, with no broken edges, and securely mounted; use the correct tool to tighten and loosen.

SHARP STUFF

The edge of a chisel should be straight and with an even bevel along its blade. A misshapen edge will never make a flat surface so it is necessary to get sharpening down to a fine art. Pick up a chisel for inspection, grip it loosely in the fist and run the thumb gently along the end (not good practice for wood chisels as there will be blood!). What counts here is that you should feel an edge keen enough that, with more pressure, it *could* draw blood – if it is not felt then the chisel needs work before it can be used for anything more than a wedge or lid lifter.

FIRE-SHARPENED CHISELS

Start off on the steel wheel of the grinder: stand comfortably, wrist on the rest and, holding the chisel firmly at the correct angle, run it squarely once across the rotating wheel observing the edge. A stream of sparks (wear eye protection!) will flow out; keep this constant for an even chamfer across the blade. Rotate the chisel and do the other side – move evenly and constantly, but not too fast as the chisel can judder and score the edge; working steadily will get a better finish.

Fire-sharpened bullnose needs to be rotated around the line of the shaft to get an even edge.

TEMPER

It is possible to lose the temper on the blade, rendering the chisel useless. If it gets too hot from the grinding process, keep a container of water in which to habitually dip the chisel during grinding; mount the container securely near at hand and keep it filled but remember the safety implications of having water near an electrical tool. Next, progress to the sharpening stone, nowadays a diamond-impregnated pad. Grip the chisel in the fist (similar to holding it for work) consciously keeping it at the right angle, and use the other hand to push it along, applying constant downward pressure, in smooth sweeps. Repeat this a couple of times and inspect the result: the chamfer should be of an even width, dead flat and of an even non-reflective tone (this will take longer to accomplish if the grinding was rough).

When not in use, keep chisels in a toolroll to protect the edges.

BIAS

When the other side is finished, sight down the cutting edge to check for perfection and repeat till the blessed state is attained; everybody will tend, in the beginning, to sharpen with a bias to one side which will result in a narrowing of the chamfer, which can be corrected by twisting the wrist to apply extra pressure to the narrower end whilst sharpening.

Use two hands for sharpening; one to keep the correct angle on the edge bevel, the other to press down and guide.

TUNGSTEN CARBIDE

TCT edges are ground in on the appropriate wheel (usually coloured green) of the grinder in the following manner: assume the same stance and grip and slowly move the edge along the wheel – there will not be flying sparks like the fire-sharpened, just a small starburst of colour at the edge to indicate where the abrading is taking place, so keep this constant as before. The material is much harder and will not grind as easily as steel – do not force it as this will result in uneven wear and make it too hot; the crystalline structure means that rapid changes in temperature will result in the edge crumbling away, so getting it hot and then plunging into the water is not a good idea. After each pass on the wheel place the edge into your palm to check the temperature; if it is too hot to bear, stop grinding until it cools down – never quench. The honing is the same as other chisels.

No shooting sparks with TCT, so keep a good angle and look closely to check the grind is happening evenly at the cutting edge.

BULLNOSE

Keeping a good edge on shaped chisels takes some practice; often a good think is needed about how it can be achieved and how to control the application. Take the usual stance and hold the bullnose with one side ready to touch the wheel. Grind it on the wheel by rotating the chisel around the shaft. The position of the chisel stays the same and by this method the blade can be ground in a controlled manner; sharpening aggressively can change the shape of the blade and reduce the amount of cutting material. On the sharpening block hold it as before and describe an 'S' while pushing forward along the length of the block, rotating the chisel so that all the edge comes in contact at some point during the process and is sharpened. Do not move the chisel shaft in a pendulum manner to rotate the blade on the grinder or block.

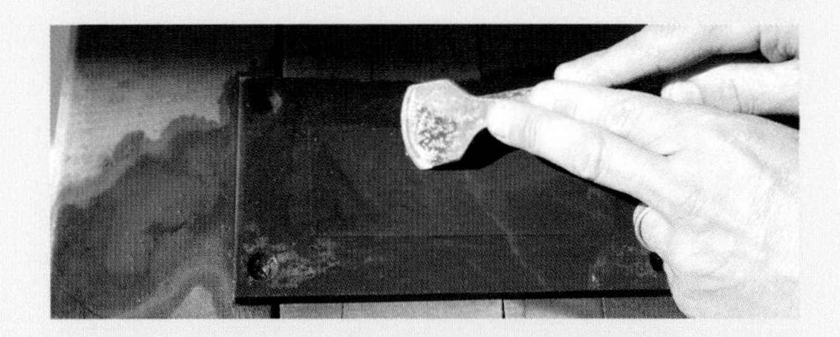

Bullnose sharpening describes a figure of eight on the stone while rotating the chisel along the shaft to ensure the whole edge is honed. To make sure you have the technique, cover the blade edge with Tippex and have a try. The Tippex should be completely and evenly removed; practise until it is.

OLD OR NEW

If you insist on using a traditional oil- or waterstone, the abrasion from sharpening will start to wear the stone unevenly, making true sharpening difficult, so it is necessary to maintain a dead flat surface on both types of sharpening stone. They can be ground flat by rubbing against a similar or harder stone; if the wear is too much get your local sawyer to cut a new face on it. Or simply buy a diamond sharpening pad of a quality that will last years; mine is old but shows no sign of wear and will last for many years to come. Dividing the cost by usage over a lifetime will show a saving over buying cheaper sharpening stone – and it is a lot easier to transport, use and look after!

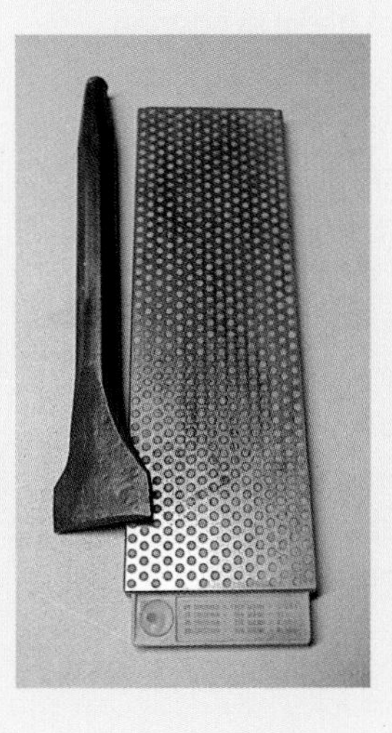

A modern diamond sharpening pad (a solid flat metal plate with coating on either side) is expensive but will last for many years and end up more economical than cheaper alternatives; this fifteen-year-old example has cost the same as a bottle of wine a year so far.

- Use the correct cutting disc for the material to be worked on.
- Always use the correct spanners/keys for changing and tightening blades/bits.
- Be sure all appropriate guards are in place, secure and working.
- Never use any accessory except those specifically supplied or recommended by the manufacturer. They should be described in the tool's instruction manual.
- Do regulations stipulate voltage when using on site? Check with the main contractor.
- Does it need to have a test certificate?

Getting Ready

- Before plugging in, ensure that power is off at the wall and the tool's switch is in the off position.
- Always have enough slack in the cable.
- Extension leads need to be in good condition; be aware they can overheat if coiled during heavy power use.
- Keep the cable out of water and protect all fittings from rain by fashioning covers from plastic bottles, etc.; use a circuit breaker plug to prevent electrocution.

Using the Tool

- Work to be cut should be secure and the cutting process must not damage other material.
- There must be no obstructions to the line being cut and enough room for the tool pass.
- *Do not* use hands or feet to hold material in position.
- Be calm; do not rush what you are doing; stay focused; do not let anything distract you.
- Do not overreach; stay firmly planted on both feet.
- Always keep a firm grip with both hands; use side handles if they are there and aim to keep control.
- Never use power tools if you are tired, sick, distracted, or under the influence of drugs or alcohol.
- Turn off and unplug the tool before you make any adjustments or change accessories.
- Allow the machine to come to a stop by its own volition – do not jam it into scaffold boards, etc., to slow it down.
- Unplug, clean and store the tool in a safe, dry place when you have finished using it.

Street site in Aix-en-Provence; just about everything is wrong here. Do not be tempted to take shortcuts just because it is a quick and simple job; fate laughs at ineptitude.

Helpful to Know

- Power tools are rated by watts; if the tool is not powerful enough it may burn out. Is there a strong smell of burning or excessive heat being generated in the tool? Stop and consider if the power of the tool is too low.
- Check that all shafts are not loose and that they rotate freely; spin and listen for grinding noises or catching as this may mean bearings are worn.
- Check that seals over bearings are not damaged, as dust and fragments could enter and wear out moving parts.
- Clean dust out of switches and lightly spray with lubricating/water-dispersing oil.

Maintaining Cables

One of the most annoying issues when using power tools is the propensity for their (power) cables to get tangled, become trip hazards or be damaged due to lying across the workspace. Wrapping them up is time-consuming and the repeated bending of them near the handle increases the chance of the wire inside breaking, causing short circuits or interrupting power. A useful solution is to change the original cable for a heavy-duty pre-coiled cable; this is available in any length, so reach can be improved, and when unplugged it can be retracted out of the way. The investment of getting this done by a qualified electrician will be paid back by the drop in problems affecting working time, a neater workshop and tools that can be readily identified (cable is available in many colours) on site.

A sensible modification for power tools is to fit them with pre-coiled heavy-duty power cables; this makes for fewer tangles on site and is beefier than the factory-fitted option. The metal blade is a plasterer's joint rule, useful in many ways – cutting back mortar repairs, a form for mitres, setting out (45 degree angle at its end), a mask for pointing and so on for the last twenty years – get one!

Money Back

Always keep the warranties of power tools purchased; they will have a tough life and will often give up the ghost far sooner than those in other trades. So be prepared to return to the store and, if possible, collect a replacement.

LINING UP HOLES

Getting holes for dowels to match up in flat slab or square blocks is fairly straightforward; when the surface and object are shaped it can be a bit more difficult. Get the pieces and place them together to check they fall into place easily. If they do not, put tape at the edge and draw a pencil line, or two, across so you know where they sit. Drill pilot holes in one piece and place a ball of clay at the hole mouth. Push the two pieces together then ease them apart; where the clay has marked the undrilled side is where to drill. If the clay does not make a mark, do it again but colour the top of the clay with a felt-tip pen.

Marks on undrilled side.

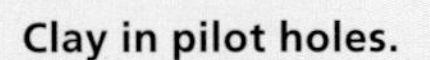

Clay in pilot holes.

Holes and dowels all ready.

CHAPTER FIVE

LIMES AND MORTARS

Throughout this book the keyword, apart from *stone*, will be *lime*; so it behoves us to get a little technical and further our practical understanding of this intriguing material.

The major components in construction, such as earth, stone, brick and wood, have fairly straightforward roles and, happily for the average building expert, can be readily identified and a correct repair or replacement can be effected (unless unfeeling technology is blindly applied or a rogue builder happens along, then things can go awry). For a long time, when construction materials echoed the environment the buildings were conceived in, the role of lime in construction and maintenance was a foregone conclusion. Unfortunately the rise of Ordinary Portland cement (OPC) meant that limes were seen as unnecessarily complicated and ineffective in comparison to the performance of cements, a problem we will all encounter. But historic buildings erected before the twentieth century will have (original) mortars and often surface finishes based on a lime binder; therefore understanding and respect, if not reverence, for this material is essential.

A rich limewash finishes off this work to the Town Hall in Cirencester. Note the similarity of this to those ancient buildings carved into rock at Petra or Lalibela; it is always a prestigious statement when a building looks as if it were carved from a single stone, and authority has always been about prestige.

Lime in Building

Lime can be found in every aspect of the fabric of historic buildings; it was used to consolidate the ground after compression by spraying with limewater, when forming foundations in many types of floor, and through all the usual masonry and wall applications up to the setting of slate and tile roofs, and on into painting and decorative schemes.

OPPOSITE PAGE:
A Roman tower in Aix-en-Provence stands as an enduring testimony to the wisdom of building with lime.

Misleading Evidence

It is obviously handy to have a render (known as torching) to the underside of a roof. This would provide insulation, fill

English Heritage extensively restored Bowhill, Exeter, an earthen and stone medieval manor house (*c.* 1425), during the last days of its Direct Employed Labour Force, which signalled the end of a trained, experienced organization that was not wholly driven by time and profit. Mortars were mixed on site and the historic evidence of past repairs kept as part of the fabric history; reroofing was controversial due to the setting of the tiles in lime mortar, as mentioned elsewhere.

Once this stone was cleaned it had lost the golden patina of the rest of the building so was given a coloured shelter coat to match in. This was achieved by mixing a range of tones applied by sprayer to fade in shades and build up colour without brush strokes.

gaps and provide suitable surfaces to decorate; and in some historic buildings it was part of the mason's work to bring them into use. Now, torching is a separate component from the tiles or slates; it is usually set on lath, thus leaving a gap between it and the tiles, which gives good ventilation and prevents the inside from being contaminated by the outside elements. The issue here is that some advocated that the tiles had to be completely set on mortar, right up to the drip edge. This was based upon (scant) evidence of lime mortars attached to roof tiles found in archaeological surveys, examples which in all probability had been make-do repairs to refix stone tiles and slate in certain buildings, the lime being used as a mastic when the nail or peg was inaccessible. This caused problems.

Why is it Wrong?

The reason for using a flat impervious surface (the tile) as a waterproofing on a sloping roof is that when the rain flows down it is to some extent drawn under the leading edge of the tile but then drips off onto the next tile (because of the big overlap and the empty space), finally being ejected at the eaves without entering the building. It is therefore not rocket science to realize that, if the roof covering is comprised of tiles on a continuous bed of porous mortar flush with their drip

A church porch roofed with lovely weathered slabs of local stone has been completely wrecked by sealing up all the edges with an ugly, hard cement mortar; obviously this was done to stop the ingress of water, but now it will trap in moisture. The timbers will decay, the roof will collapse and because they are all attached to each other with this mortar most of the tiles will break.

edges, moisture will be drawn up the underside of the tile and into the building, creating damp patches and rotting out the wooden pegging or nails holding them in place.

THE SMEATON PROJECT

This was undertaken by English Heritage as a research project into various types of lime and pozzolanic additives to determine the performance of historic and contemporary mortars as binders in varying proportions. The conclusions, garnered from years of state-funded research and summarized below, are so simplistic as to be pitiful – any competent artisan could give the same answer after a moment's thought.

- The materials in use today are not always the same as those used by the builders whose work is to be conserved and preserved, even if they are called by the same name.
- Materials are not always being used in the same way.
- The performance requirements of a repair mortar may be significantly different from those used in conventional modern contexts.

Corfe Castle. We built these steps to blend in with the ruins and were being progressive with the use of hydraulic limes. Unbeknown to us, the manufacturer added OPC to level out fluctuations in strength, so technically this was cementitious work and caused a huge hoo-ha in the approach to hydraulic lime use.

Binders

When building with masonry or other block component it is standard practice to have a plastic material (that sets or hardens) as a mortar infill between the individual stones. This material has several functions depending on its composition, application, the fabric material and the method of construction. Note: there is no mention of adhesion.

Binders other than lime (or cement) that have been used in the construction and repair of historic buildings include gypsum, which in its raw state is a form of alabaster, a stone used in many church monuments and decoration (be aware it is water-soluble and should be cleaned accordingly). Gypsum can be manufactured more easily than lime due to the lower heat needed in its production; this made it an attractive option when wood was scarce as a fuel. It has the disadvantage of being more soluble than lime (though this is an advantage in the cleaning of polluted stone surfaces) and

Town house, Bath, built of the local stone and lime. This city exemplifies the beauty of integrated design and a common material. There are issues though; the stone itself is not very durable and is affected badly by pollution, and this house underwent extensive repair with much new stone to the parapet and chimneystack. Though the ashlar looks solid, it is only cut flat a certain way in and most blocks could be termed wedge-shaped, with rubble infill supporting it. Grouting in a replacement window entablature (first floor, far right), the liquid grout actually ran down through the fabric and into the ground floor room at skirting level; this was due to the good-looking but poorly made façade.

The render for this church tower was mixed with dung and applied literally by hand; this possibly accounts for the weird circular staining patterns that subsequently appeared.

is not an option to be considered for exterior work, though exceptions occur. Clay was sometimes used as a mortar by itself or with little aggregate; it needed very little processing to use, and was often a natural choice in rural vernacular situations, sometimes being used to construct the whole building as earthen, cob or adobe (half the world still lives in earthen structures). Clay mortars should not be replaced with lime.

Other unusual mortar binders have included white lead, linseed oil and dung, and there could be more waiting to be discovered.

Portland cement (named because it was said to resemble Portland limestone when set – see natural cements below) started to be used in 1824, when Joseph Aspdin patented it. The development of Ordinary Portland cement (OPC) can be attributed to the time of the building of the Eddystone lighthouse (by John Smeaton in 1756), where lime with a relatively high clay content was found to harden under water, as opposed to high calcium lime that needs to be in contact with the atmosphere for carbonation and could not perform satisfactorily in water-logged applications. This was in effect the re-discovery of natural or Roman cements – actually eminently hydraulic limes that had been used in antiquity to great effect; OPC is a synthetic equivalent of these.

Fast-setting and easy to use, OPC is much evident in many of the worst stone repairs. Here in Shaldon, Devon, there has been no attempt to make it look neat, let alone reflect the style of the window – truly hideous.

THE ISOLATION LAYER

When restoring an artefact that has had surface finishes applied that need to be preserved, it is best if they can be presented in a way that gives future work an indication that there is a definite limit to the extent of that work, so future investigators can identify the work that has been added. This is the isolation layer; with this in place a date, presuming records are kept, can be given to any work after finding it during investigation.

LEFT: Royal coat of arms, Newlyn East, Cornwall. This was modelled in a lime plaster, and was promoted as being bare of colour originally. This idea does not seem right as when heraldic devices were plain the colours would be indicated by specific hatching for each shade (this being a standard practice in heraldry), yet here there is none. A thoughtful art teacher in the 1970s added the glum colours evident here, which are quite unappealing. CENTRE: After microscopic examination of the paint in cross-section to determine previous schemes, it was decided that to remove the paint would be impractical and damaging, so a new scheme based on the findings was proposed and after a good clean an isolating layer of soluble acrylic was applied, allowing the story to begin again. RIGHT: Afterwards the vibrant colours (heraldry is all about show, so no muted tints here) were applied and the golden parts covered with gold leaf. This was quite a change from what everyone was used to, yet the finished result more than holds its own in terms of aesthetics and veracity.

Commercially produced hydraulic lime mortar comes in sealed bags, tailored to specific demands, will have a definite ratio of water to dry given and is easy to mix; this ensures a fully consistent mortar – as long as it is used correctly.

Lime in Mortar

For the purposes of this book, we will consider historic replacement mortars and renders based solely on lime. So what is this panacea for the support of our built heritage?

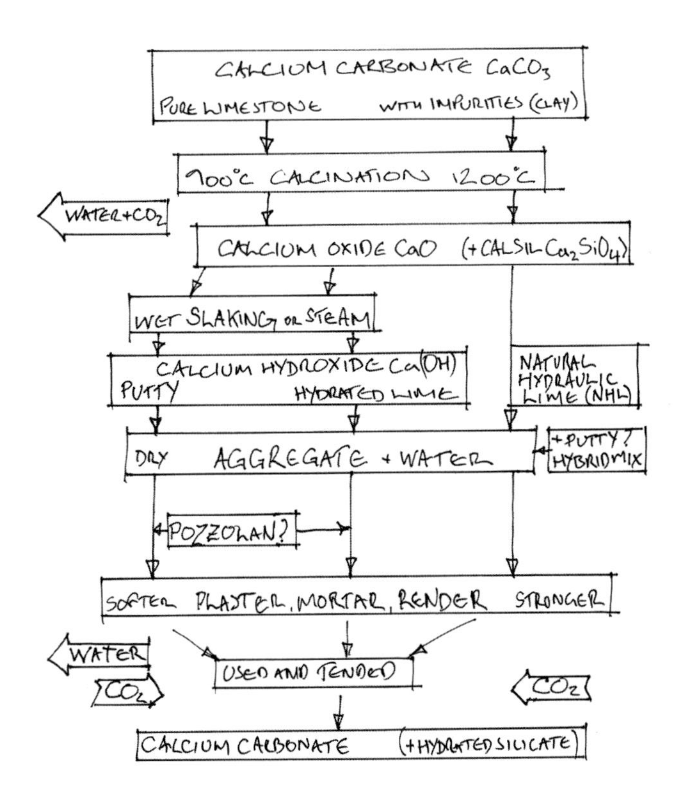

The lime process. Note that it is not a cycle, as the mineralogy is the same at start and finish, but not the material! The left side is for non-hydraulic and the right for hydraulic; with eminently hydraulic lime the levels of carbonate are very low and come closer to OPC in a natural manner.

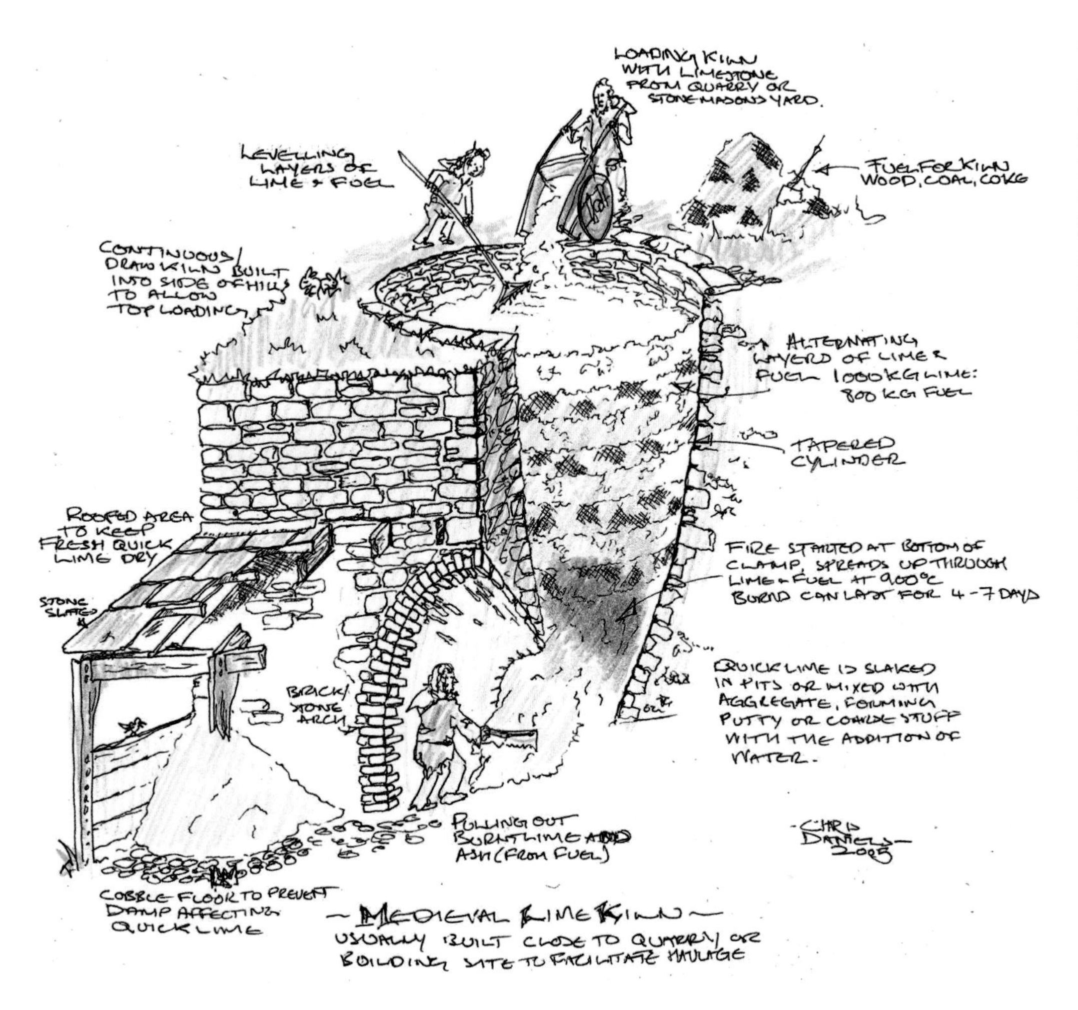

This is a cutaway drawing of a typical draw kiln. The crushed limestone and fuel (coke or charcoal) is built in layers and fired, the temperature at the bottom calcining the stone and converting it to quicklime. It would be pulled out of the bottom, under cover in case it rained, and more raw materials would be tipped in at the top for a long (often days) burn. (Photo: DWG)

Lime Production

There are three main building limes all readily available nowadays:

> two of these are hydraulic or non-hydraulic, supplied as a hydrated, fine white or grey powder in dry sacks. The other is lime putty, a glutinous white paste, supplied in tubs with a protective covering of water, which is always non-hydraulic.

The Lime Cycle

All three are produced in the same way up to a certain point. Lime (calcium hydroxide or hydrate) is produced from the action of adding water to quicklime or lump-lime (both terms for calcium oxide), which is the product of burning or calcination of the raw material(s) that contain a percentage of calcium. The raw material used is calcium carbonate, which is found (for the gourmands amongst us) in its purest form as eggshells and oyster shells; obviously relying on these would restrict its availability to the back doors of restaurants, so for any usable quantity stone is used, sometimes coral or marble but most often the easiest to procure, chalk and limestone.

Dolomitic limestones, even though they can contain a high proportion of magnesium, are also used for producing a high-quality oxide for lime production.

The Process

First the raw material has to be calcined or heated to a red heat (between 950° and 1070°C) to drive out carbon dioxide and moisture. Originally this would have been in a simple clay clamp (basically a mud pile with a fire in it), often producing unburnt stone or over-fired lime due to uneven temperatures. More practical and reliable was the flare or draw kiln where crushed stone and the fuel would be loaded in layers and fired; here quicklime was removed from the bottom, as more raw materials could be layered in at the top.

The construction of the kiln and the quality of the fuels used meant that lime produced in these ways could have significant inclusions, such as clay, unburned fuel, etc., that could affect its performance.

Local Colour

Most vernacular building would use materials sourced very close to the structure's site; and so the builders had to take what they could get without too much quality control. These impurities and the use of local materials created singular products, presenting us with difficulties when reproducing a mortar to match original work. It may be that up to 75 per cent of these locally produced limes could have had some hydraulic properties, from inclusions.

Nowadays the cost of production compared to consumption precludes small-scale indigenous lime burning, so most available limes come from sources located at great distance (in the comparative terms of historic building), so we start off with a material that can be quite different from the original in terms of purity and inclusions. Manufactured in gas- or oil-fired rotary kilns, this modern lime is a standardized material that will often have to be modified before it can be considered for use in conservation work. The majority of lime produced in this country in industrial quantities is for use in the steel industry; as this needs to be modified extensively for heritage work, it is often pragmatic to use hydraulic lime for specific applications.

Movement

One of the benefits of using lime is the lack of immediate strength, allowing the (inevitable) settlement of a new building to be absorbed by the mortar rather than stressing the masonry. This, in my view, questions the obsession with high compressive strength mortars; I am unaware of any masonry failure due to the mortar collapsing under compression.

Remains of a kiln next to the quarries on Ham Hill. These were for burning all the waste from working the stone and also the overburden (which probably would have contained clays, often producing some hydraulic mortars). These types of kilns are dotted all over the country, showing how the local vernacular was based on local materials.

Shapwick, Dorset. This church is a fine example of local building, with flint and ironstone courses, larger Purbeck blocks (possibly looted from Corfe Castle), imported Bathstone for the dressed work, Purbeck stone tiles and later brick buttresses. All attest to using what is available. There were many lime sources around at the time, some producing hydraulic, some air-limes, but all used in exactly the right place. The building had settled and the resulting large fracture has been recently pointed up with lime.

Ratings

The NHL rating is a modern system of classification that relies on the minimum compressive strength achieved (in a specific time and under controlled environment) by a mortar using lime and a standardized aggregate. NHL stands for naturally hydraulic lime and *not* non-hydraulic lime! This still causes confusion for some when a lime mortar is specified, as there can be many variations of this material. Probably the same can also be said for cement-based mortars; at one end of the lime–mortar spectrum the distinction between the two binders is blurred, and rightly so, considering that developments in cements (during the eighteenth and nineteenth centuries) resulted in many cement-based historic renders – a point that is very often overlooked.

Non-Hydraulic Lime

This is used in a putty-like state and consists of (non-hydraulic) lime that has been slaked with greater quantities of water (40–50 per cent) than used for bagged hydrated lime. It is common practice in countries that have unbroken masonry traditions to soak hydrated lime to produce putty; here in the UK there is always a purist element that reckons it lacks the cohesion and workability of pure putty and insists on the expensive route. Whichever method is used the product must be left to mature for a significant period, the longer the better – the Roman architect Vitruvius insisted lime should mature for at least three years. Lime putty bought nowadays tends to be thirty days matured, whereas when we used to slake, it would be done once or twice a year to get the best material.

A large bag of happily maturing lime. The glutinous paste is sticky and makes lovely mortar; it is almost rich enough to spread on a scone with jam! (Photo: Rose of Jericho)

Curing

Non-hydraulic limes set slowly by a process of carbonation drawn from the atmosphere and weak crystallization of the soluble components. It loses significant amounts of moisture as it dries, and without decent aggregate this results in much shrinkage, without the development of any great strength. Carbonation will not take place below 4°C, but performance and setting can be improved by the addition of reactive hydraulic compounds.

These limes are usually categorized from 'fat' to 'lean', with fat lime or high calcium lime having the best workability when slaked and matured. Lean limes contain less calcium, having impurities that are usually non-hydraulic and occasional impurities that cause it to perform as a feeble to semi-hydraulic lime (see below).

Controlled curing of a mortar repair, using a bespoke 'greenhouse' containing a damp sanitary towel to keep the moisture level correct; to be effective it must be sealed in.

This is often seen: cotton wool is placed over new mortar to cure, a method that is pretty much useless as the water held in it is pushed out when it is applied. This causes lime to wash out, while the high surface area will accelerate evaporation. It needs constant attention if used outside and really should be relegated back to its original use in museum work.

Hydraulic limes, cured by a crystalline set, are classified by an NHL (naturally hydraulic lime) number that gives the lowest compressive strength of a mixture of the NHL and sand in Newtons/mm^2. This classification is not exact in real terms as strength can vary widely relying on a multitude of outside factors: sand type, porosity of fabric, temperature, relative humidity, wind speed, time of day, protection, tools used, ratios of mix materials, drying control, compaction, hours of sunshine, aspect of building, shape of fill, depth of fill and so on.

In the depths of winter, with heavy frost on the ground, the work here should have much better protection – or possibly should have been done in the warmer months. The new block of stone has been fixed and beaten into place, resulting in the fracture. It also needs a chamfer to match the existing; hopefully this will be worked afterwards.

Core capping work, placed twenty years ago, was using a mix of non-hydraulic and hydraulic lime, cured properly and has lasted well. There is much debate about using 'hybrid' mixes due to them failing. Personal experience shows that it is all about the application and tending – many cannot cope with the differing properties of the two and get it wrong. I spent a week beneath this block on my stomach, cleaning out debris (and some archaeological finds) before walling up the entrance.

Feeble to Semi-Hydraulic Lime, NHL2–NHL3.5

These can have a smaller proportion of free lime (calcium or magnesium oxide) whilst containing a higher proportion of hydraulic components derived from the clay content of the limestone the stronger they get.

With this lime, essentially calcium oxide and calcium silicate means slaking is similar to non-hydraulic lime in the action of water on the oxide, but the slaking process will take slightly longer, due to the time needed for the hydration of the silicate.

This hydration is the setting process, and is coupled with the carbonation of the hydroxide (as in a fat lime); as the process is quicker there is less shrinkage due to moisture loss.

Eminently Hydraulic Lime, NHL5

The chemical composition of an eminently hydraulic lime is similar to Portland cement, with a small to tiny proportion of free lime that can aid workability. The hydraulic components are reactive clay compounds of silica, alumna and water; these may also be present in non-hydraulic lime but as nothing happens when they are not compounded and thus reactive, they are just classed as impurities.

These limes consist mainly of silicates, and in extreme cases the amount of calcium oxide may be negligible; this makes the slaking reaction unnoticeable, but hydration, and subsequent setting, takes place rapidly in a few days.

The calcination process of hydraulic limes is important as the formation of clinker (so named for the sound it makes when hitting the floor) can occur at higher temperatures; clinker is relatively inert unless it is reduced to a powder by grinding. If (like Portland cement) the clinker is ground extremely finely, then natural cement is produced.

Hydraulic or 'Natural' Cements

Parker's 'Roman' cement (patented in 1796) and other natural cements were produced by calcining stone with a high clay content to a greater temperature, causing them to sinter or clinker; the results would be subsequently ground to a fine powder. These were produced from shale known as 'cement stone' found naturally along with hydraulic limestone but containing a higher proportion of clay, or from naturally occurring clay limestone conglomerate lumps.

Today we have 'Prompt', a natural cement that sets within minutes of mixing – good fun for the uninitiated!

Natural cement, readily available, should only be used with awareness and experience – it is not for the faint-hearted. Retarding agents, in the little bottles, can help control the beast.

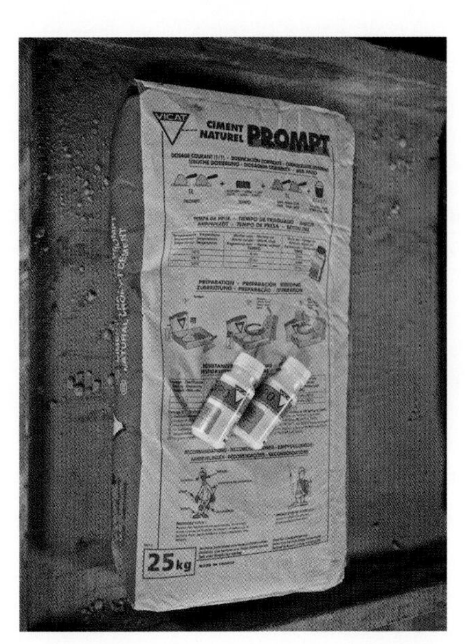

Casting a backing support layer for a crest using natural cement and LECA, with reinforcing inserted for attachment to the wall. The plaster case was to hold the pieces in place while vacuum consolidation took place, and will come off easily.

OPC

Modern (Ordinary) Portland cement is a very different material from that patented by Aspdin in 1824 (today's OPC does not bear any resemblance to Portland stone). The original was in reality pozzolanic cement produced by adding clay to slaked lime and burning the mixture at a temperature too low to produce complete vitrification; this was then ground, thus making it different from hydraulic lime. By 1845 the advantage of complete sintering was realized (and thus Ordinary Portland cement) and this became standard practice.

Working with Lime

Slaking the Safe and Cheap Way

If the need is for a non-hydraulic lime, and putty is not available or unnecessarily pricey, then it must be made at the workshop a long time before it is needed, so go to a local builders' merchant and buy a bag of hydrated lime, half-fill a big tub with water and add the lime, stirring well till it is all soaked and a stodgy mass; it should be a ratio of about 1:2

water to lime. Do not worry if it seems too runny as the excess water will eventually separate out as clear limewater on top of the putty.

Not So Safe but a More Fun Method

It is possible to buy quicklime from the manufacturer in sealed tubs or tough plastic sacks; this is a highly volatile substance and caustic to boot, so great care should be taken when slaking.

If this is going into a pit, get an old metal bath (or two if more is needed) and set it securely on blocks with the plughole overhanging the pit. Take out the plughole assembly and make a wooden bung attached to a pole that can be lifted out without getting your hands into the mix.

Now put on the protective clothing, gloves and eye/head protection and fill the bath to a depth of about 15–25cm, then commence sprinkling the quicklime evenly onto the water. When an amount has gone in and is soaking up the water, drag it through with a rake or hoe until the whole is mixed completely. As the slaking goes on there will be lots of steam, the mixture will get very hot and small plukes (pimples) in the mix will fizzle and pop; depending on the

Slaking lime in the factory; the bag is set on top of the rotary mill mixer and fed in at the same time as the water is added. Evenly slaked, it comes out ready for tucking away to mature. (Photo: Rose of Jericho)

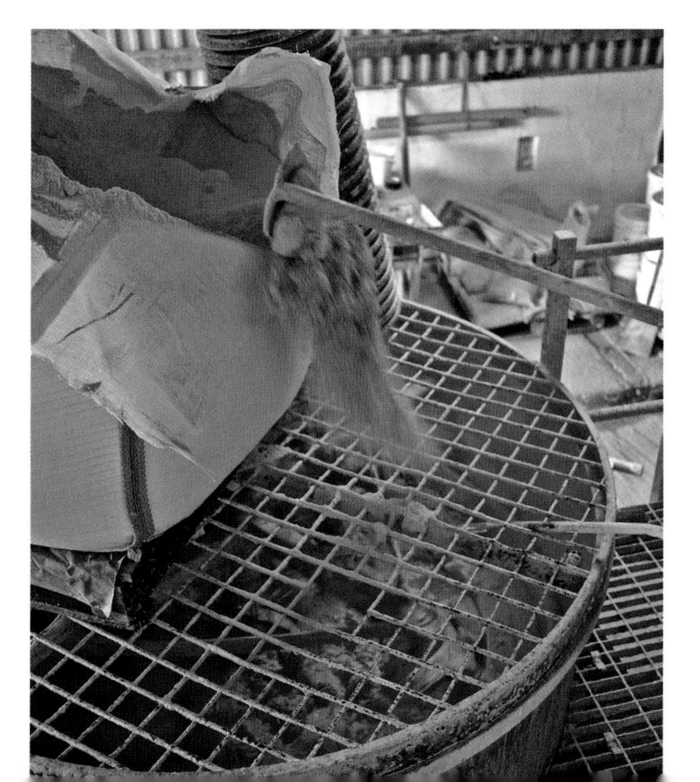

WAR STORIES

LEARNING THE HARD WAY

When the use of lime started in earnest, it was difficult to get hold of quicklime, except direct from the quarry, where it was made for the steel industry. As that was a massive industry, there was no provision for small-scale users like us, so it was necessary to buy it by the tonne; this was loaded into the pickup from a vast hopper more used to filling lorries of over 30 tonnes. This resulted in a huge pile of highly hydrophilic material sitting in the back of the truck, forty miles from the workshop. Unfortunately, as this was the west of England, on one trip a sunny day soon turned wet, and thus it was, after a surprise cloudburst, that the slaking process got under way while I was on my way home, creating a huge cloud of steam and exploding chunks of quicklime far and wide – enough to cause severe traffic disruption and threat of prosecution!

So now quicklime is only available to small-scale users in sealed containers or sacks – probably the best thing really.

Introducing local historical group at Corfe to the joys of lime, and for this wearing appropriate clothes.

THE IDIOT WAY

Slaking quicklime produces lashings of heat and, if using large lumps rather than powder, explosions, which can throw the caustic stuff about. Working as a callow youth with my chums, we used to carry out this process at the back of the workshops, wearing only shorts and sunglasses, the idea being that if we were splashed by the lime we could hose it off straightaway; scars from my lime burns are still visible today.

Nowadays it is advisable to wear full protective gear, attiring oneself almost as if to go to the moon but, again, *this is far safer.*

freshness and type the initial slake will take up to half an hour. Keep raking, and once it is a sloppy enough mix to flow (with a push maybe) through the plughole, take the bung out and empty it into the pit. Two baths will allow one to stand and slake, while the other is filled. The slaking will continue in the pit for some time. Do not forget to have a secure waterproof lid for the pit; and if you do not slake all the quicklime that day, make sure it is covered and dry.

Practical Matters

Non-hydraulic lime putty, once slaked, should be stored in a covered pit and left to mature for a minimum of a month. This ensures complete slaking and will produce a better material (we have already noted Vitruvius' recommendation that lime for mortars be three years old at least).

Semi-hydraulic lime run to putty should be used within a week; after this it may deteriorate due to the hydraulic set occurring.

As slaking is speeded up by using warm or hot water, there is a theory that slaking pits on site would have been used as urinals by the workforce; this could explain the traces of urine and salts in old mortar. The warm urine would have started up the process effectively and been helpful perhaps when using inefficiently burnt stone; a useful side effect would be that the pollution caused by a large transient workforce would be kept to a minimum. As the properties of lime are antiseptic and as it was known that the addition of fats and animal protein could be beneficial to the waterproofing and binding qualities of mortar, the lime pit may also have been the recipient of food and animal waste. There also may have been a degree of saponification between the lime and the fats, which is echoed by the modern use of detergents to cause air entrainment for durability and frost resistance.

Mortar

In ashlar and block construction where the structural integrity of the masonry is due to the fine joints and the manner in which the stones are laid, rather than the strength of the mortar, the mortar acts as a cushion to spread loading evenly across each joint and bed. It also functions as an expansion joint, allowing movement in the fabric without incurring failure that can be the result of too-rigid materials.

This tower in Exeter had lost its structural stability, and had steel bands wrapped around it to prevent collapse. Limewash (of a much pinker hue than it appears in the photo) was used to finish it.

When wall construction becomes coarser, as in rubblework, the mortar will play a greater part in ensuring the strength of the masonry. By keying into irregularities and preventing slide, it can form a homogeneous filling that will resist lateral movement – all without the qualities of the hard, unyielding, cementitious mortars that actually stick the stones together.

Old and New

The major difference between the use of mortar in present-day construction and traditional building (this term is used to cover methods and materials prior to the widespread use of cementitious mortars) is that the mortar today plays more of a rigorous structural role in the construction, as opposed to being a gap-filler. This ignores one of the main tenets of building – that in the majority of applications the mortar should be softer than the stone. This concept is based on the premise that buildings should not resist the inevitable as accumulated weight settles, but actually move with it; this has been utilized

Calcite staining indicates that the mortar had not cured sufficiently before there was enough moisture in the fabric to wash it out in solution.

Carved brick repairs using mortar had the advantage that the crushed brick aggregate improved the set immensely.

for thousands of years. It is also helpful to the longevity of a building if the mortar decomposes before the stone by forming a conduit for waterborne salts and allowing them to collect in the easily replaceable mortar. This is an indicator as to the reasons for the general survival of so many traditional buildings.

Using porous gap-fillers may seem like a contradiction in terms, and the fact that the mortar will degrade over time could be considered a minus point, but these are performance qualities that may have the effect of prolonging the life and integrity of the masonry.

What to Use

It is still not widely accepted that non-hydraulic lime can be an effective binder for mortars and renders without the addition of some cement; it is seen to achieve only a gentle set over long periods of time, and in some cases has never hardened. This can allow the building to settle and 'find' its best position without creating undue stresses; it also allows seasonal movement to occur. On the down side it can inhibit building high in one session if used as more than a fine jointing mortar, and on external work there is a definite season that has to be awaited when the weather conditions will not adversely affect the performance of the mortar. It is therefore essential to be aware of the individual requirements of each application and of the way that the mortar should and will perform to meet these criteria. The use of hydraulic limes of varying eminence is one way of preparing mortars; the other more historically relevant way is to modify the lime by additives.

Additives

Creating artificial hydraulic lime has often been attempted in times past by adding clay to chalk or limestone, but sporadic results due to difficulties getting the burn correct and eventually the progress in OPC caused the further exploration (on an industrial scale) of this to be sidelined; however, it is known that in various historical periods, right back to Classical Rome, additions were made to lime, subsequent to its calcination, which would improve setting and strength. Generically termed pozzolans, these additives would be of a clayey nature that had been subjected to intense heat, such as volcanic ash (natural) or fired earth (artificial). They are usually a siliceous, aluminous or ferruginous material which, while not cementitious on its own, reacts with calcium hydroxide to create an artificial hydraulic lime.

Incidentally, these are supplied as fine powder, so their inclusion in a mortar mix necessitates a rebalance of the fines level, where the additive is now upping the percentage.

Pozzolans

True pozzolana is a reactive volcanic ash from a region of Italy around Mount Vesuvius, near the town of Pozzuoli, and was used by the Romans to produce concrete as used in the dome of the Pantheon and the Pont du Gard aqueduct at Nîmes; it is still used extensively today in the Italian building industry. Trass is another reactive volcanic ash sourced from the Rhine valley region and used widely across Europe, while tuff is an igneous rock that can be ground down for its pozzolanic properties. Soils and sands containing silicas or aluminas, usually derived from the decomposition of igneous rocks, can also be natural pozzolans to a lesser degree; if using an unusual aggregate, it may be worthwhile to check it for hydraulicity using the test below.

Artificial Pozzolans

Clay burnt at about 850°C and finely ground is the main artificial pozzolan; obviously low-fired bricks and tiles can be crushed down for this as well, adding a bit of colour to the mix as a side effect. Pulverized fuel ash (PFA or fly ash) from burning pulverized coal reacts with lime to form good cement (grout for masonry core consolidation uses PFA as its main constituent). Fumed silica is a very lightweight filler material that also has useful attributes in this realm. Ground blast furnace slag is used in the production of special cement, as it is pozzolanic; so it is another to add to the list.

Brick powder or high-temperature insulation (HTI) are products of fireclays and can be obtained as a fine powder, rather than undertaking the tedious and hazardous task of crushing on site; most of these will also improve the mortar's resistance to frost, though PFA and HTI will not set in cold weather. Ground clinker has been occasionally used as a sand substitute and imparted hydraulic qualities to the mortars.

The important thing is to know the materials you are using, so if embarking on the modified mortar route, start making samples using all the additives and limes you can lay hands on; use a standard mortar for the model, and note the differences the additives impart (obviously this includes the colouring effect) – be creative.

Aggregates

An aggregate is used in mortar for its mechanical and cohesive properties; it also aids workability, reduces shrinkage and provides a cheap bulking material – these properties depend on the type and quantity of aggregate used.

The mastery of mortars and concrete has produced some spectacular architecture throughout the world.

The Maison Carrée in Nîmes, built over 2,000 years ago, is the best-preserved Roman temple in Europe. A study of the dimensions would show how the entirety relies on definite proportions and ratios to work.

SPOT THE POZZOLAN

As mentioned, it is good to know if an aggregate or filler has pozzolanic properties, so for any unknowns run the following test:

Put 0.5gm of the material and 0.3gm of slaked lime in a test tube and add 22ml of distilled water. Shake well at twelve-hour intervals for a week. At the end of the week make up the same again and compare after shaking both. If the volume of sediment in the first is greater than the new mix, the material is pozzolanic. To gauge the degree of this, compare the volume increase to that of a known pozzolan that has undergone the same test.

The volume increase is caused by the reaction of silica and alumina in the pozzolan with the lime solution, producing alumino-silicates of calcium, which are colloidal; they occupy more volume and also settle at a slower rate, to give a good measure for comparison.

Sand (sharp, soft or a mixture) is the usual aggregate material employed for the majority of building applications. Using sharp sand will give strength due to angularity of its grains, which lock together in such a manner as to discourage movement. Soft sand, composed of rounded grains that slide together easily, though producing a weaker mortar, imparts greater workability to the mix and will aid the compaction and gap-filling properties of the mortar.

Sand itself is made up of the ground particles of many minerals, shells and fossils, a product of the deposition and erosive factors of the local geology and climate. It is fairly common for mortars to use sand indigenous to the area, so it often happens that a specific sand may have been worked out or become unobtainable. Original sands may have been used in a contaminated form, such as sea sand, and this can be one of the originators for salt damage.

When good-quality sand is compacted together properly it will lock into place, demonstrated here by sand sculptures that use no binder apart from water; the effect will also be found in a good mortar, with the added advantage of a binder. (Photo: SandWorld)

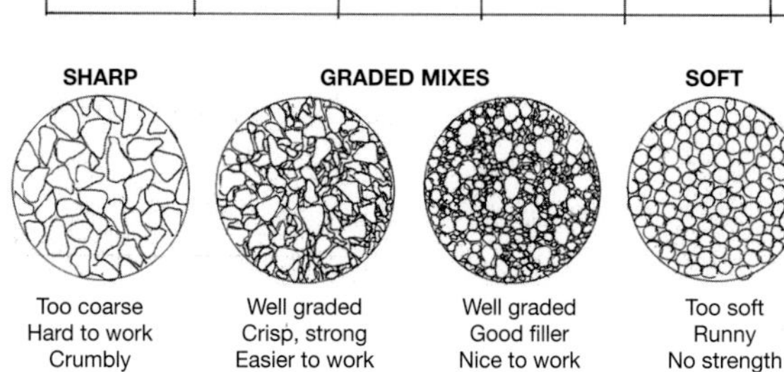

Different shapes and grades of sand.

Sand samples ready to be matched up for mortar or analysis. (Photo: Rose of Jericho)

Bits and Bobs

Other materials used as aggregates or present in mortars can be:

- Stone dust or chips
- Marble dust
- Gravel
- Ash
- Chalk
- Kiln slag
- Shell
- Brick and tile dust

These may have been included deliberately, which may be apparent through observation of their distribution through the sample (this is why it is good practice when obtaining mortar samples for analysis to choose from various parts of the section under review). Or their use could be incidental by using any granular material to hand (kiln waste, fire cinders, etc.) to bulk out the mortar (not an unusual practice by most builders), resulting in unique inclusions that also could be pozzolanic and thus alter the performance of the mortar. Here it generates a quandary: if the mortar performance was accidently altered by random additions, should the replacement mortar copy this or should it replicate the original design for the mortar?

The far western end of Chesil Beach in Dorset running from Portland to Lyme has, due to the sea currents, become naturally graded; locals are said to know which part of the beach they are on by the size of the pebbles and gravel.

Grading

The grading of sand used is important. Lime is a paste and will surround the separate grains when well mixed, but it cannot give support, so all voids will have to be filled by the aggregate itself.

Sand to be used as an aggregate should be well graded and washed to prevent inclusion of salts. Soft sand should have a maximum of 5 per cent silt; too much fine silt will make a weak mortar and can affect porosity, so stick your hands in the pile and work the sand through your fingers. Take them out and wipe on a white cloth – if the cloth ends up dirty then there is probably too much fine stuff/clay in there.

WENTWORTH GRADE SCALE FOR SIZE OF AGGREGATE

Grade limits to		
	256mm	Boulder
256mm	64mm	Cobble
64mm	4mm	Pebble
4mm	2mm	Granule
2mm	1mm	Very coarse sand
1mm	500 microns	Coarse sand
500 microns	250 microns	Medium sand
250 microns	125 microns	Fine sand
125 microns	63 microns	Very fine sand
63 microns	4 microns	Silt
<4 microns		Clay

Bulking

The volume of sand can be affected dramatically by bulking. This is where each grain of sand is surrounded by a film of water preventing it settling; the sand is wet enough to stick together (comparatively) and can occupy more volume than dry sand. More water will cause the particles to slide together and reduce the volume; completely saturated sand occupies the same volume as dry sand – it is this effect that is used here to test for bulking.

Sand is stored in bays, so obviously it will tend to get wet and possibly bulked. If only buying a small amount, get a shovel and mix the sand about first to evenly distribute the grains, then bag it.

Get a clean glass jar or beaker and scoop up the sand to completely fill it, without tamping it down. Now gently pour water into the jar and gently stir with a stick (in technical language this is known as a non-metallic agitator), adding as much water as possible. Leave it to stand and the water to clear (fines will cloud the water). Look at the resulting height of the sand in the jar; if it was bulked it will now be lower than before the water was added.

Use a ruler to measure the original height and the finished height and express the difference as a percentage of volume. The answer tells you how much allowance should be made when batching the aggregate.

Mixing

As stated, non-hydraulic mortar can be supplied ready mixed, though this will still need some preparation prior to use.

Due to the sticky consistency of lime putty it is hard to obtain good mixing using the standard rotary drum cement mixer; usable results can be achieved by placing large stones (hard smooth stone that will not break and crumble) in the mixer and letting them 'pound' the mix. Strong, large-capacity industrial mixers are best here as the pounding is considerable and goes on for longer than a normal mixing period; alternatively, portable paddle mixers, of a decent power, are a good investment.

Old iron tracks on Lyme Regis beach were used to transport blue lias to make hydraulic lime.

Pre-Mix

Traditional methods were to mix the aggregate in with the quicklime and then slake the mixture; this ensured the best possible results, with all the aggregate evenly covered – so try this out if you are doing any home slaking (it will involve a bit of maths to get the volumes correct). This also works for hydrated lime; just let it mature for a decent period then knock up again.

There are many advocates of labour-intensive methods, where the ingredients are put on a board and either chopped together with a spade or stamped in by foot (hopefully they have the sense to take off their sandals and put protective footwear on); apart from interesting demonstrations or very small jobs, this is patently inadequate and ridiculously messy for any serious amounts needed on normal sites.

Serious suppliers and users have large mills that roll the ingredients together (and also crush aggregate down to a powder if required) in quantity; they also manufacture hair plasters using this type of machine.

Once mixed, the mortar/render can be stored indefinitely in sealed or covered containers as coarse stuff that, once matured, is highly workable. The mortar should always be knocked up thoroughly prior to use to ensure maximum workability and quality. Any additives will be added to this mix and once this is done it must be used that day, preferably within four hours. If it is not used and is stored again it should be remembered that the (pozzolanic) additives cannot be relied on to impart their usual qualities.

Haired mortars need to be used fairly soon after manufacture as the hair will soften and rot in the wet mix. However, these are for plasters and renders, so not really relevant here, but it is a useful point to know. Hair does not add any strength to a mortar; its most useful role is when squeezing mortar between lath, where the hair prevents the nibs falling off.

Haired lime render has come adrift from external lath here at Bowhill in Exeter; this part of the building is new and obviously does not have the staying power of the original.

Mixing Ratios

It is generally agreed upon that the optimum ratio of lime (binder) to aggregate is in the region of 1:2–2.5. This ensures that there is maximum distribution and coverage of particles by the binder. This is based on the standard assumption that bulked sand will have up to 30 per cent of its volume empty; the lime is intended to fill these voids.

The ratios can be varied depending on the application and performance criteria required. With material other than sand used as an aggregate/pozzolan then it is necessary to tailor the proportions to obtain the best mix.

Analysis of Mortar

Sometimes it may be necessary to analyse the original mortar to find the ratios of binder (aggregate and the types of sand and inclusions). The analysis will be obtained firstly from experienced observation of the mortar style, its texture, colour, aggregate type and sizes, inclusions and historical setting. This takes practice and experience but it is surprising how much information is there.

The second method will be laboratory analysis. This should give a breakdown of the binder ratio and the grading percentages of the aggregates. The mineral content of the aggregate will be ascertained by magnified or microscopic examination, or, in big jobs, by X-ray diffraction. Obviously the cost of the analysis rises with the intensity of the investigation.

As there can be unavoidable discrepancies due to past practices of reusing old mortar in new mix, and using calcium carbonate minerals in the aggregate (both of which will affect the testing of the binder ratio), the results of laboratory analysis will still have to be interpreted by an experienced conservator in order to utilize the findings in the recreation of the mortar or to design a suitable replacement.

Most firms that carry out analysis charge to analyse only one sample. This may not be indicative of the whole building, unless chosen with absolute certainty; it is prudent to consider this when pricing for external analysis.

This is a personal innovation, a plaster cast that can be pointed up with a mortar to show how it will look in two widths; the aggregate used is stored in the case to allow for quick inspection. This is cast from a silicone mould of a piece of stone with the joints cut in, a cheap but practical archive system.

HOW MUCH LIME?

To find out the volume of lime needed, it is necessary to find out the volume of voids in the aggregate.

Place a large pile of sand on a baking tray and dry this out by cooking it in the oven. (Incidentally, to find out how much water you have paid the supplier for, weigh the sample accurately before and after cooking – it might be an interesting result.)

Fill a litre container with the dried sand and tap the sides to settle the sand as closely as possible, topping it up until it is completely full. Then, using a full gradated beaker, start pouring water into the sand until the water reaches the top of the sand. Check out how much water was used, express this as a percentage of the whole, and *voilà*! This is how much lime is needed to create a perfect mortar.

A penetrometer is used to test the density of a mortar by dropping a weighted point into a tub of the mix. This is useful, when on major projects, for tracking the quality of materials.

CHAPTER SIX

REPOINTING

The iconic Royal Crescent, Bath, one of the world's architectural gems; built with fine ashlar work of tight flush pointed joints, it gives the impression of monolithic construction. This idea that a building was carved from one piece indicated great wealth and importance, harking back to a time of god-like rulers able to command legions of artisans to realize their dreams and convert mountain to palace.

The image of beauty and unity of Bath's hallowed terraces does not quite hold up once one gets behind the façade. The original purchasers were in effect buying a lovely front with nothing behind it; it was up to them to get a builder in and create the house behind. Like most towns, this would be driven by one-upmanship and a complete lack of co-operation between neighbours; the result is a mish-mash of design and quality, with no two houses the same. Cost cutting, after all the real money was spent on appearance, has produced some real horrors.

When constructing a building or monument from stone, it is usually from a drawing, or when repairing, by following an example, so the surface finish of the stone and the type of jointing will be laid out in the specifications. What is usually missing, though, is information on the type of pointing required, which is one of the most important aspects of a façade or structure. Good pointing can give the whole a pleasing aspect, while bad pointing can destroy any aesthetic qualities of the masonry. The spaces between stones, bricks and other components of a masonry structure are filled with

OPPOSITE PAGE:
St Catherine's, Abbotsbury. The stonework of rich honey limestone glowing in the evening sun is supported physically and aesthetically by the unobtrusive lime mortar.

this plastic material that allows stones to be safely laid without pressure points. Good pointing will prevent the ingress of water on the surface and play a sacrificial role in the dispersal of moisture carrying harmful salts. Here it is our role to replace or replicate this material from a binder and aggregate that is (hopefully) less durable than the masonry; we know that even our own work will eventually break down and need to be replaced.

Understanding how to produce a good mortar, and the successful application of it, is one of the most important skills to have as a stonemason or restorer; as with all this work, it requires skill, attitude and intelligence.

Think About It

To include the word 'intelligent' when talking about the packing of mortar into a gap between stones may seem like overstatement, but it is necessary to use your brain as much as your hands in this or any other situation when building or repairing. Being competent in pointing and repointing of walls is a fundamental requirement in the repair and maintenance of historic buildings or masonry structures – it is not good enough to have a half-hearted approach to what seems to be a boring job; it makes a difference.

The sheer loveliness of this wall is from the playful movement of the stones and the galleted mortar; almost an instinctive build with no pretensions, any repair would demand skill and judgement, as it would be all too easy to ruin it.

USE YOUR EYES

Look at the wall; what is noticeable? Usually it is that the condition of the mortar is not what it should be, or that there is a mish-mash of styles and materials, so have a look around and identify original work, as this is what should be replicated, or at least honoured by complementary work. Fashions change and the finishing of pointing is no exception; how should the mortar be finished? In my youth the trend for any wall not ashlar was 'ancient monument style', where the mortar was cut back from the surface to 'present' the stones – this was purely aesthetic and stemmed from the romantic notion of ruins, harking back to the lack of mortar between stone due to loss.

It is truer to tradition to apply flush pointing on rubble work, brought to the surface and level with the stones; this can be extended to run over the stones and then be rubbed over with burlap to let the high spots 'peek' through. This is almost plastering and, as far as historical veracity goes, is probably the most pertinent. The issue with this is that it can, for the first decade or so, look quite messy to the owner and so the trend is to go for the neater recessed version.

The pointing here is a prime example of unintelligent decision and slack standards; rusticated pillars on the Town Hall at Bridport have been overfilled to the point where they change the appearance of the stone, or at least they would if it had been carried out to all the perpendicular joints as well – this is not only unpleasant but slack. Using unskilled labourers to do this is bad enough, but how it passed inspection is a mystery; the mess and the hideous mortar repair should have been enough to condemn it, even by those unaware of historic detailing.

Old mortar gradually eroding away, keeping up its end of the bargain by providing a large surface area to volume ratio as the pores enlarge, essential for the wicking away of moisture.

The pointing here on this end of Glastonbury Abbey offends the eye, and rightly so – the mortar was made with a plastic binder, luckily only a brief fashion in the 1970s; and we all know about 1970s fashion . . .

Is It Really Dead?

Aside from empty joints, there will be two instances where repointing is necessary; firstly when the original mortar is breaking down to such a point where it does not do its job anymore, identifiable as excessive crumbling away or being unattached to the stone. But be aware that mortar can *appear* incredibly moth-eaten and crumble away with ease, yet still be doing its job, which is to allow moisture to get out of the structure. Eroding mortar has a lot of surface area (pores) which can enhance its function of moisture wicking; do not get carried away digging out full joints just because it is easy to do.

Cementitious Mortar

The other situation is when there is an incorrect mortar used, causing (or with potential to cause) damage and problems within the masonry or to the stone itself. This usually consists of a hard impervious mortar based on OPC, termed 'cementitious' in any literature or report, as the binder. (Organic plastics have been used in mortars, and are so wrong they will never be mentioned again.)

So, ignoring the natural degradation of original/correct mortar, what we have is a bogeyman of a material that has been, and still is, used extensively to make mortars that are fast-curing, sticky and almost waterproof – all properties that appeal to the uninitiated when faced with a structure that needs to have gaps filled quickly (that is, profitably) and to prevent the ingress of water into the masonry. Seen in countless examples of repair to historic walls, this notion is readily

Cement mortar has been smeared over the decaying arrises of sandstone ashlar; the original joints were very fine and should have been pointed as such with a decent mortar. Now the stone is eroding around the joints and once again the repair is causing more problems.

undermined as the mortar becomes detached from the stone, cracks (because it is too brittle) and forces the unavoidable moisture to get out somewhere else – with disastrous results. OPC also puts soluble salts into the fabric as the moisture leaves it, which is another reason to loathe this stuff, though it can be used beneficially in some circumstances. It is never going to be simple in this game; much prejudice for and against materials is based on myth, misinformation or just plain ignorance.

Cement Terms

There are two definitions of the word 'cement' that will be used throughout this book. The first is the verb, where it is used to denote the attachment of material to itself or others. The second is as a noun that describes a manufactured material widely used in the building industry for the production of mortar and as a binder, originally called Ordinary Portland cement; while this can be also a white cement, for ease of understanding, whenever cement as a material is mentioned it will be termed OPC.

A wall on the brink of destruction, the cement mortar has proved it will always shrink, the subsequent cracks pulling in moisture and encouraging the growth of invasive plant life.

Basic Considerations

Correct materials and methods, when used wisely, can perform the jobs of preserving, protecting and waterproofing, among others. With regard to the context and style, the right mortar can be utilized to enhance the appearance of the walling material or render it unobtrusive where necessary. Unfortunately this work is too often regarded as mundane and not accorded the concentration required to achieve worthy results. Carried out in a slapdash manner or without due thought to the materials, the methods or the fabric it is applied to, it can be destructive in the worst situation, and useless even at its best.

Pointing? Mortar repair? Render? Whatever this could be styled, it does not fit well with the wall. There is some texture but joints have been jumped, and the colour is too pale for the quantity.

Before You Start

The use of the correct material is essential to the effectiveness of the mortar; the general consensus is like for like, unless removing inappropriate mortar when a match to the original is to be desired. When a building has a soft mortar as its original, then a soft mortar must be used when repointing. If it appears to be only hard pointing in the joints, it is essential to check that this does not hide an original mortar.

Aggregates

Mortars use aggregates to bulk out expensive material, impart strength and reduce shrinkage. The aggregates in historic mortars are wide and varied, and can include crushed brick, shell, grit, kiln slag, ash, stone fragments and dust as well as the more usual sand (the term we shall use in this book for aggregates in general); these will all affect the performance.

As mentioned before, 'sharp' and 'soft' are definitions given to sand that denote the shape of the grains. Sharp sand has angular grains that can 'lock' together, giving a strength and cohesiveness to the mortar. A purely sharp mix can be difficult to work with and be weakly cemented if the sizes are all similar. Soft sand grains are more rounded, which helps them slide around more easily, producing a more workable mix; though they are easier to apply they can produce little cohesive strength. If these are not well graded then the mortar can shrink excessively.

The best ratio is one where the binder fills the voids within the sand, and to achieve this the sand should be well graded, where it has varying proportions of different sizes, dependent on the intended use.

Cleanliness, not Fines

The sand should be free of clay and organic inclusions, washed with fresh water and dry before mixing. Too much fines (clay or dust) will cause a weak mortar and should be easily less than 10 per cent. To determine this, place a scoop of the sand into a glass beaker (or old jar), filling it halfway. Pour enough water to come almost to the top of the container and stir the sand vigorously for a minute or two (half a teaspoon of salt will aid the settling out). As the sand settles heavier grains will be at the bottom and the fines will form a layer at the top. Use a rule to measure the thickness of the layer and calculate it as a percentage of the total height of the sand – if it is too much do not use the sand. In the builders' supply yard this may be a bit difficult, so there just rub the sand in your hands or on a white apron; if it balls up or leaves a residue then it is too dirty and should not be used.

A sieving machine, for checking the grading of aggregate. The dry sand is put in the top case and the whole tower is vibrated, shaking the sand through progressively finer mesh. The catch in each sieve is weighed and given as a percentage.

A good colour match and texture means nothing if the tending is negligent. Here frost has destroyed this pointing; it might have survived if it was summer, but in the middle of winter with no protection after application it was doomed.

MORTAR ANALYSIS

The aim is to find out what the aggregate was and, hopefully, the proportion of this to binder. English Heritage has done some research into historic mortars and their aggregates. The resulting publication has illustrations of various made-up mortars; unfortunately this is only printed in black and white and is pretty useless for comparison. (The reply I had, when mentioning this, was that anyone in this field should have a Munsell Soil Colour reference book to use the charts; as this costs in the region of £200 it may be prohibitive to buy, and it does not help any one wanting to learn more.)

The advantage of keeping good-quality samples as archives is that they can be used for getting a visual match to a mortar sample. (Photo: Rose of Jericho)

QUICK LOOK

Crush the sample down and give it a good swirl around to wash the aggregate, and then see if the type of grains can be identified. A search of information from quarries can often yield pictures of their product, though if there are local supplies of sands try these first.

WORKSHOP ANALYSIS

The next stage is to carry out a disaggregation analysis, which can be carried out in the workshop without too much trouble.

- Take a sample from the wall of 50–100g, place it in a sealed container and take it to your test site. Weigh this sample as it came out of the wall (this will give some information on how damp it was), powder it down then dry it in an oven at 110°C for twenty-four hours.
- Weigh this to 0.001g to gain W_1.
- In a glass beaker, pre-wet the sample with a little distilled water then cover with hydrochloric acid (HCl, sometimes called Spirit of Salts), and watch as the binder is dissolved.
- Dry a filter paper; weigh it to get W_2, then place in a funnel over a glass beaker.
- To make sure the dissolving has completed add a couple of drops of HCl to the sample and stir.
- Add distilled water and stir with a glass rod to suspend the fines, then carefully pour through the filter; the solids will stay at the bottom of the beaker. Repeat until the water in the beaker is clear.
- Dry the filter paper in the oven at 50°C for two hours. When dry weigh for W_3. The weight of the fines is: $W_3 - W_2 = W_4$.
- Swirl the aggregate in the beaker in distilled water and pour off the water, then dry the aggregate in the oven at 110°C for twenty-four hours. Weigh for W_5.

The restorer's sweetshop! Here in the Rose of Jericho office are myriad types of sand, and examples of mortars all made up and ready to avail yourself of; organized and well laid out, companies such as this are making everyone's job easier and more relevant, but just as useful should be personal samples from previous work. (Photo: Rose of Jericho)

RESULT

Percentage of aggregate = $(W_5 / W_1) \times 100 = A$
Percentage of fines = $(W_4 / W_1) \times 100 = F$
Percentage of binder = $100 - (A + F)$

CAVEAT EMPTOR

Mortar analysis is often required for restoration projects, so samples are sent off, analysed intensely (expensively), and the results used to specify the new mixes. Problems arise as many old mortars used limestone as aggregate and this could be made up of even older mortar. The presence of calcareous elements other than the binder will slew the proportion analysis, as they get dissolved along with the binder. Silicates (from clay) as impurities may not be distinguishable from hydraulic compounds.

In reality, to practically determine the mortar mix, look at the original, crush it down and identify the aggregate (check local sands for a match), work out how it was applied and what levels of porosity it needs, then design a suitable replacement. Matching all the petrologic and mineral nuances of a historic mortar is archaeologically sound but expensive, and could be seen as bringing inappropriate levels of technology to an area where materials were measured by the shovelful, depended on the skills and mood of a medieval labourer, and if they were caught short then any old stuff (which may have modified the setting properties) could be chucked in.

Mortar make-up kit. Here is the set-up to weigh, sieve and mix mortar samples. These are then packed in the brass mould and cured; basically it is cookery for building!

Avoid overuse of pigments to get a colour; instead concentrate efforts on using suitable coloured aggregates and stonedusts. When assessing a colour be artistic in approach and look at the background build-up of hues. It is easy to buy a red-coloured sand, but some reds are based on a soft brown while others are hard purple; always look at colours, test panels and pigments in natural light.

Non-Hydraulic Mortar

For fine joints, interiors and low exposure external zones, non-hydraulic lime putty that has matured for at least a month after slaking can be used; containing up to 45 per cent water, no extra water is needed when the appropriate aggregate is added. Alternatively, hydrated lime can be used, prepared by mixing dry with the aggregate before the addition of water and then left to mature prior to mixing; this produces a really well-distributed mix. Non-hydraulic lime mortar will keep indefinitely (maturing all the time) if covered in storage. Always knock up this mortar before use; the plasticity increases the more it is worked prior to application.

It is claimed by some that mortar made with hydrated lime does not possess the same degree of 'workability' as that made with lime putty. This is purely subjective; matured it will produce a first-rate workable material.

Built of granite, Castle Drogo has continuously suffered from moisture getting into the fabric and has been the subject of many fruitless investigations and intervention. Note the ugly pointing and, more significantly, the white stains; this is calcite leaching and indicates the mortars used have not cured before moisture has been able to wash out the binder.

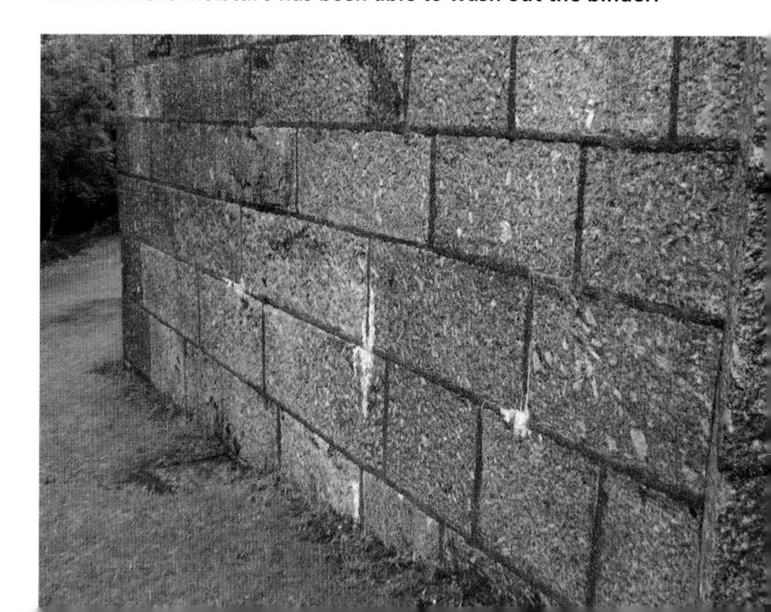

Strong, coarse pointing well matched to the stonework here on Hardy's Monument. Note how the new stones have rich tooled surfaces, essential to bring texture and life, as well as helping to blend in with original.

Hydraulic Mortar

It is accepted now that most exterior mortars, whether for pointing or rendering (surface covering), should have some level of hydraulicity. This can range from adding reactive components to non-hydraulic to using strong hydraulic limes; all should be site-specific and not just on a whim.

Removing Old Mortar

Joints must be cleaned out thoroughly, removing old, useless or too hard mortars from previous work. Chisels and scrapers of various designs can be used to cut out joints with care, avoiding damage to the arrises of stones or dislocation (loose stones that may or do come adrift must be noted and returned to position). Generally disc cutters should not be used, as in the wrong hands they can widen or overcut, though a fine diamond disc used properly can cut and relieve hard cement pointing to facilitate removal with minimum damage to arrises. There are power tools that use oscillating blades to clean out joints; care must be taken with these, especially if the stone is softer than the mortar, as the blade can slide off the mortar and cut into the stone.

SAFETY

Lime is an alkali and will cause extreme discomfort and possible permanent damage if eyes are contaminated; if this occurs wash with fresh water or suitable eyewash (it is compulsory to have this on any site where this, or any work is undertaken) and seek medical aid *immediately*.

Contact with lime can also affect the skin, causing drying and irritation. It is advisable to wear gloves and use a suitable barrier cream. *Always* wash your hands thoroughly *before* going to the toilet, as sensitive skin is particularly at risk.

Chest tomb conservation included lifting the ledger, removing ferrous cramps and refixing on lime mortar; localized repointing and highlighting the remaining inscription letters using acrylic paint tidies up the appearance.

Raking Out and Removal

Soft mortar can be dragged out using old saw blades; these could be anything from hacksaws up to those from bow saws (remember to make a handle and protect your hands). If there are lots of wide joints, a screw or nail head projecting from a hard wood handle will do the job well. If this is going to be a lot of your work it may be wise to make some tools specifically for this task. Always pull the tool, as this makes controlling the depth easier; pushing will be harder work, resulting in jamming and spalled edges. If the joint is too tight to get a fine blade in then it will be difficult to get a mortar in.

Quirk and bespoke claw chisels (here TCT for toughness) are the primary tools for raking out any mortar that cannot be scraped out. Their design allows them to get in tight joints and apply the force in a way that pushes the material out or breaks it up efficiently.

A selection of raking-out blades; also a hand blower which is pretty but not powerful enough for any serious work. Plastic packers of various thicknesses are essential to have around, or collect lead sheet scraps, cut to size and flattened with a mallet.

Denser Mortars

With a sharp pointing tool or chisel of the right width cut into the mortar at almost right angles, and the waste should always be pushed into an empty space. Using a mallet rather than a hammer will lessen fatigue and keep a better rhythm. Do not jam the tool in the joint – if necessary use a narrow chisel or be prepared to grind raking chisels narrower. Remember that they need to be thinner behind the blade to well past the joint depth.

Raking out old mortar to large joints in rubble stonework while wearing gloves is sensible here. The chance of flying bits would question the absence of eye protection and nowadays safety helmets are becoming compulsory especially for public works. Note the all-purpose building tools here; the labourer's bolster has a plastic guard for protection from maladroit hammer use striking the hand. It is good practice to introduce stonemasonry hammers and the correct chisels on sites dealing with work of this nature; with practice they can augment the proficiency level of even the lowest hand.

OPC Mortar

This will exist in three states, the first being where it is solidly adhered to both sides of the joint. In this case, question whether it is necessary to remove it, as doing so will probably damage the stone. If it must go, the best way is to relieve the mortar by using a small disc cutter with a fine blade. Get into a comfortable position and carefully track the blade along the centre of the mortar incrementally to prevent digging and slipping, until a channel is cut to a good depth. It will aid stability if you support your arms by resting them against the wall; if your overalls catch on the surface a knee protector on your elbow can help. Also be aware of the closeness of your head to the cutter – PPE is a must. Now with a sharp TCT chisel, firmly place the blade parallel with the joint near the arris of the stone and hit sharply to pitch the mortar off the stone. Minimize damage to the arris – try not to jam into the gap between stone and mortar. In most cases the dead mortar shards will fall out; those that jam in will need to be broken up with a TCT chisel in the joint.

Cleaning out wide rubble stonework joints with a rotary burr; used with skill, tools like this can speed up the work and as there is no impact from hammer and chisel, there is less chance of loose stones.

For OPC mortar stuck to one stone with a gap to the other stone, use the same chiselling technique as above. When the mortar is detached from both sides, use a TCT chisel with its blade at right angles to the joint and sharply strike to fracture the mortar into shards. Do not dig in, as jamming will happen.

The prepared joint should be back to solid material and cut in square (not tapered). It may be necessary to wedge stones in place if the raking is of enough depth to dislodge the stone; slate tapped in between lead is ideal for this – try not to use wood.

Washing

Loose dust and remains of mortar can be washed out with a light pressure hose, also wetting the wall, an essential task to reduce 'suck' on the new mortar, unless there is a lot to wash out, when it may dry out again. Washing out will produce slurry running down the wall, and in worst cases stains – be ready to wash the mess off the whole wall as it happens. If there is a problem with dirty water running onto paths, etc., then roll hessian at the bottom as a soak-up. Compressed air

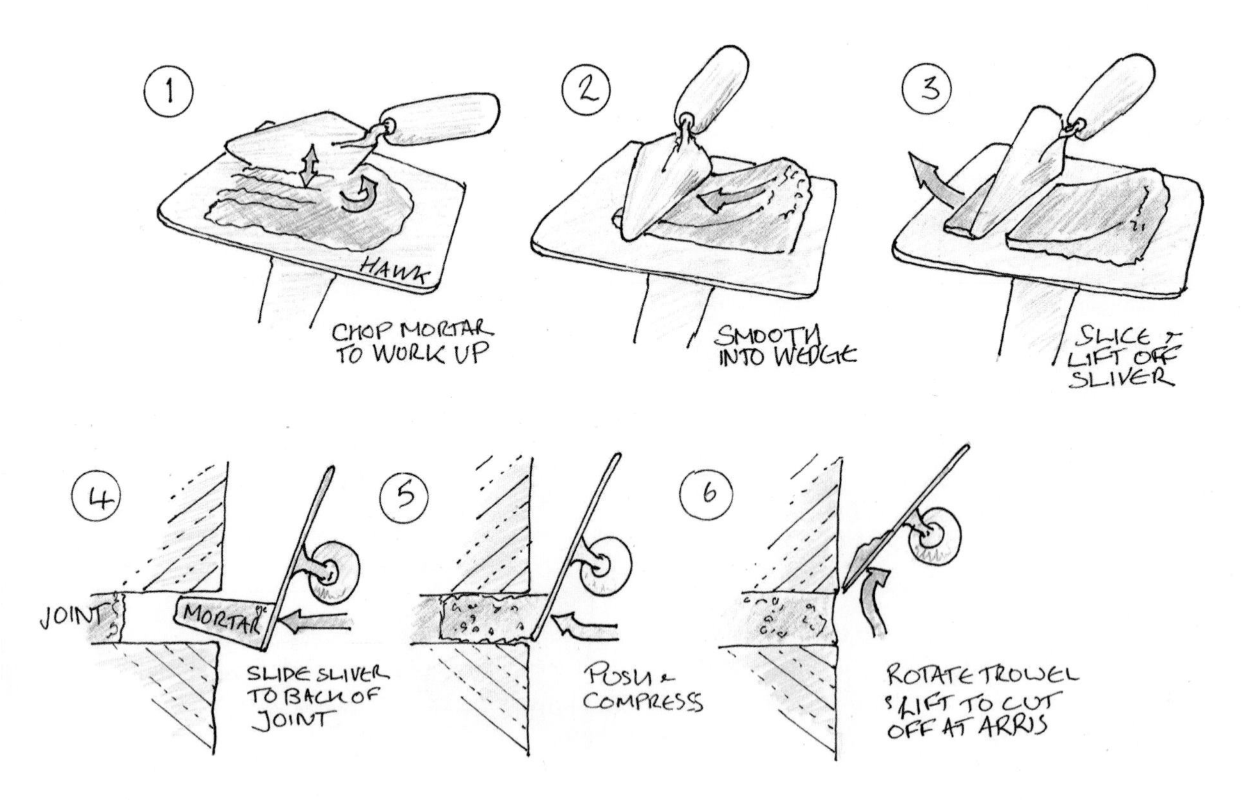

Pointing techniques: the trick is to have the mortar exactly right in consistency for working up and sticking to the trowel. It is also important to place and compress the mortar without getting any smeared onto the face of the stone (or brick).

CATCH TRAY

Working at height on scaffold lifts will lead to a lot of mess falling down, so make a tray from some spare board with a fence to prevent stuff rolling off; this can be pushed up against the wall to collect the debris. Walls can often be bordered by soil or vegetation which may need protection; using a tray also reduces cleaning up, and when pointing, it will also prevent the inevitable lumps of mortar falling down and possibly staining stone or paving below. This device is also handy because small tools can be put down without fear of them falling between the boards.

Mortar droppings landing on stone have the potential, should a cloudburst occur, for giving the wall an impromptu and undesirable limewash.

is a fast and effective way to get rid of the debris, but it can be chaotic and messy; wear eye protection and ensure the dust is not settling somewhere that is a problem (cars, windows, etc.). Brushing out of joints is not really effective for cleaning and will tend, even if actually getting to the back of the joint, to taper the opening or deposit fines at the back of the joint.

Filling

In ideal circumstances, before any mortar is touched, start the wetting process three days before filling by intermittent mist spraying that gradually makes the stone and remaining mortar completely moist. Use a mist attachment on a hose or, better still, a paint sprayer with compressed air as this produces fine mist easily absorbed by the masonry. A hand-held sprayer is not enough for the task without lots of work. It is best to do as much wetting up as possible to get

A decent sprayer is essential for all lime work. This one is over eighteen years old and though it cost a lot it is still less than the total of economy versions that would be needed for the same period. Here it is used indoors to wet up a fireplace prior to repointing.

Here is what happens if the raking out, wetting up and tending are not all done properly; the mortar dries out too quickly and becomes white and crispy. This mortar never stood a chance – it cannot be saved, only removed and done properly.

the best performance from the new pointing; consider that when originally building the wall the mortar was as one and the whole thing would be completely moist, so there would be no 'suction' from dry spots. Spray the surface to the point of droplets appearing, stopping before the water starts to run down; once the water is absorbed, repeat. Cover during the wetting-up process to retain the moisture level.

Knocking Up

When happy that there is enough moisture in the wall, it is time to ready the mortar. If a hydraulic mix is going in this should be mixed now; let it have a good long pounding in the mixer and then load enough into a sturdy bucket to keep going for a while. Non-hydraulic ready-made mortar should be taken out and any (pozzolanic) additives thoroughly mixed in now – do not add water; put onto a spot-board and with a big trowel or shovel give it a good working through (knocking up).

Please resist the urge to pound it by foot. The result will be not satisfactory and there will be very tenacious white boot-prints on all surfaces, with annoyingly messy rungs on ladders. Put the mix into the bucket. Mortar left on the spot-board should be pounded up into a heap (reducing surface area and drying) and covered over with a bucket or similar.

Loading the Hawk

At the workface scoop a trowelful onto the hawk, mash and pat it, working it into a nice paste with the trowel; it should be stiff enough to retain shape when you pick up a lump on the trowel. If a fine mortar is being used, remove any large aggregate grains and dispose of them – do not put them back in the bucket. Never use bare hands to manipulate mortar: it may not seem harmful at the time, but it is very hydrophilic (drying out the skin) and when you wash afterwards the friction between your fingers can become painful.

Into the Void

Shape a rectangular pat and smooth it into a taper at the farthest edge to just under the thickness of the joint. Slice off a piece wide enough to reach the back of the joint and deftly lift this so it is standing on the bottom edge of the trowel. Slide this neatly into the joint, pushing in well till the flat of the blade meets the stone. Press firmly in and rotate the trowel upwards and back so that the mortar is caught in the inside of the arris without getting onto the stone. If using a thin-bladed pointing trowel, narrower than the joint, press in as hard as possible and draw it off to the side.

Some use the hawk as a loading platform, held to the edge of the joint and sliding the mortar off into the joint; this is effective but creates some problems: sliding on the hawk will

Use the tool that works. Here a small modelling tool is filling a small joint too awkward for the standard trowels. Specialist art shops will have a stock of these, or old dentistry tools are an option.

These small scoop-type hawks are being used more often nowadays. I find they tend to be awkward and, being metal, the mortar slides around, which is not great for shaping up stiff mortar. It is purely personal choice for tools and if this does the job for you then it is the tool to use.

It was observed that this joint had actually been padded out with newspaper on a so-called restoration job by a mature stonemason (and site assessor for a government training body, no less). This, and other bodges, were brought to the attention of the authorities who still let this cowboy continue to devastate the historic wall. This is wrong at all levels and, as this is my local church, is a real bee in my bonnet.

cause sloppy mortars to clump up to thicker than the joint; and material can fall between the edge of the hawk and the stone with the potential to create mess out of sight. In addition, too much reliance on this method will cause problems for tight corners or vertical joints.

Filling the Gap

Mortar should packed into the back of deep joints, building up in layers; rough up the surface of each layer to get bite on the next. Clean washed gallets packed into the fresh mortar will fill out large spaces, reduce the amount of shrinkage and can look good – pay attention to orientation and type of stone used.

Tools should be suitable for the size and type of joint and must be handled with dexterity to ensure good compaction and prevent mortar touching the face of the stone. (If this occurs the stain must be washed off with clean water and sponges immediately, rinsing the latter frequently; once the wall face has dried it will be difficult to remove the resulting lime stain.) The filled joint should be pressed in hard with the trowel blade to compact and reduce shrinkage; this can

Here is another variation; working off the underside of a trowel to get the mortar into the joint – if it works . . .

be repeated in the first couple of hours of application for hydraulic mortars and a day or two for non-hydraulic. Whilst pointing, the wall surface and the inside of joints should not be allowed to dry out. Pump-up garden sprayers or similar can be used to keep the area of work damp while working.

Fine joints are always fiddly to fill. It takes practice to get the mortar consistency and the placing right, but once done even the finest joints can filled; just remember to sponge off all the time. Some specifications or manuals suggest a technique using tape to mask off the joint and insert mortar through the gap; there is rumour that this method was suggested as a joke that then got picked up for a book, and is now part of the lexicon for 'consultants'. It is never practical or successful, as staining becomes a problem, especially if the preparation work and tending is done incorrectly.

SPRAYER QUALITY

Hand-operated sprayers come in all sizes and prices; shop around to get the most durable, and ask yourself if it will be up to life on a building site. Small half-litre flower sprayers are easy to carry about and will be good for a day, and then will probably need replacing, so buy good quality ones that are a decent size (to reduce trips to the tap). Grit will destroy seals, especially on sprayers that pressurize by pump handles on top, rendering the sprayer useless – so add a bead of petroleum jelly to the seals on moving parts, and do not carry the sprayer by the pump-handle as this causes stretching of seals. Be prepared to dismantle and clean such sprayers regularly – or buy new. Heavy-duty back-pack sprayers with lever pumps are expensive but can be easily found second-hand and are best suited to this work; they will have high capacity, are easily carried and have well-protected seals.

Cutting Back

If the pointing is to be set back from the surface of the stone, or perhaps to have a struck weathered surface (in this case, meaning sloped to shed water), then fill the joint fully, compact it, and let it start to set. When it is firm but not dry, using a fine plasterer's or favourite trowel, trim by running the blade along just below the arris of the stone. This will leave a ridge along the centre that can be rubbed off, compacted and textured, using a smooth planed softwood rubber.

Pointed up strong with a good firm mortar, this will be left to start a set before being cut back.

A fine example of well-balanced, nicely textured pointing, evenly finished with no staining (always an issue on brickwork); the whole ensemble creates a gentle image with no pushy aspects.

Good clean coarse pointing, with a neat tile slip repair to a coursed rubble garden wall.

Good coarse haunching-style pointing, sloped to deflect the worst of the weather. Try to avoid feather edging in this situation.

Here a weathering surface of refixed copings, with new stainless cramps in, needs to shed water, so the pointing was finished by smoothing over with a spatula.

Finish

The finish is dependent on what is required visually and should not differ too much from the original style; one rule of thumb is to bring the mortar almost flush with the face to reveal the stone shape (ancient monument style).

In fine ashlar work that was traditionally flush-pointed often the arrises have degraded; filling will give uneven joint widths, so it may be necessary to cut the mortar back from the face to the true width of the joint. Avoid feather edging onto the stone, as this creates a weak point.

Occasionally mortar should be brought over the stones to give a semi-rendered effect; this is probably historically correct as this type of 'pointing' would be faster and easier to apply, but is not that popular as it takes a long while for the mortar to weather in (here meaning to lose its new, raw look). Rubbing vigorously with a swab of hessian gives a good coarse texture and compacts the surface to finish off this type of pointing.

The final texture of the pointing can be achieved in a variety of ways:

- Rubbing with wood when quite firm. This will give a broken texture that has a high surface area, which can aid the evaporation of moisture.
- Brushing with a soft brush, which will reveal some aggregate and seal small pits/cracks. Firmness of the pointing must be gauged accurately to avoid brush striping.
- Sponging back by pressing a slightly damp sponge onto the not-too-firm joint will give surface compaction and sealing. Care must be taken not to stain the stone face.

NOT IN LAYERS

An inappropriate technique widely specified for deep joints is to place the mortar in layers and let these carbonate before putting the next layer in; realistically though, carbonation of the mortar is certainly going to take longer than hoped and will not be ready if the job is to be completed in a practical and economic manner. Personal observation is that preparation, insertion and tending of the mortar is often carried out badly, so when attempting to do it many times over, they invariably fail. The other issue is the interface between the layers; as they are tamped in and the surface stiffens up, they will be less porous than the main body, and the result will be obstruction of the flow of moisture through the mortar. The outer layer will be more prone to the corrosive effects of weather and pollution, and wet areas can form inside the joints. Deep joints should be packed out to the face in one hit, as it were, to make a homogeneous mass with no weak points. Gallets can be added to reduce the volume of mortar. Correct control of humidity during the tending period will ensure that it cures thoroughly without shrinkage or dusting.

The massive amount of mortar being used for core capping here will be prone to shrinkage, especially when it is such a wet mix, unless tended extremely well. It was subsequently covered with turf (hard and soft capping, an unusual mix of techniques) and it was difficult to check out how it got on. The curious method of tamping it down is not to be recommended, as the caustic lime will corrode stitching in boots and the lime-staining footprints that result will not be welcome on ladders and floors.

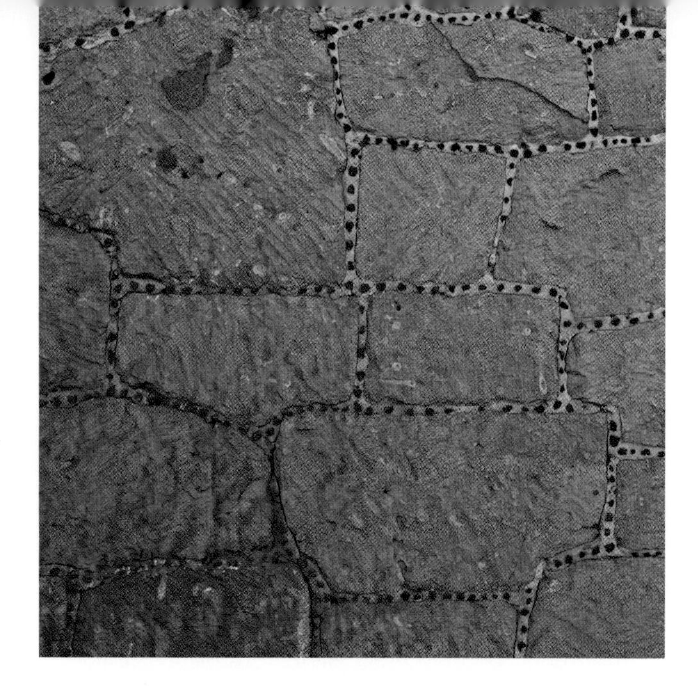

A lovely treatment of pointing in dressed-to-fit rubble stonework. This quality of work deserves to be highlighted, and the nodules of what appears to be charcoal or coal do just that.

Due to time constraints this building was repointed and continuously mist sprayed using a compressed air paint sprayer for the next twelve hours, while being allowed to heat up in the sun; the warmth and humidity produced a perfect set and subsequent cure in record time. This method is a route worth some further investigation to refine parameters and methods of control.

- Using a steel trowel will give a smooth waterproof surface, well compacted. This can be helpful in rain-washed situations.
- Full or flush pointing to coarser stonework can be brushed over with a churn brush or rubbed in with a hessian pad during the cure.

Tending to Control the Set

It is essential to prevent overly rapid drying of the mortar as this will have detrimental effects on its performance; unfortunately this happens quite often as time or a slack approach overtakes correct method. Drying can be controlled in a variety of ways – these rules also apply to mortar repairs and some extent the fixing of stone.

Mist Spraying

Working with the conditions on site, gently spray a fine mist over the area at frequent intervals. Use a portable sprayer for localized work, or a mist nozzle on a hose (useful if there is a big area) or use a paint spray-gun with compressed air. In my experience the last is the most effective; it gives an incredibly fine mist that is effectively drawn into the mortar. Do not overwet, as uncured lime will cause staining if it goes into solution and runs down the wall.

STEAMING

The perfect curing of mortar (including OPC), for analytical purposes, should take place over at least twenty-eight days, firstly at high humidity then with controlled drying – this gives the best possible set, hence the emphasis on getting the process correct on site.

Steam can be used for faster and more effective curing of cast blocks of lime concrete – this relates to the use of compressed air spraying, as the moisture produced is so much finer and insidious than normal spraying.

Covering

Protecting the work from the drying effects of wind is essential even when it is not hot (Antarctica is the driest place on earth, and definitely not warm). I have misgivings over the standard practice of draping damp hessian (burlap) over work because, unless well attended by a mist sprayer, it can actually cause a hot absorbent zone that can release the essential moisture out to the atmosphere. If it is used it should always be covered with a waterproof layer, such as tarpaulin, to build up humidity and keep it so. Plastic sheeting on its own is good on windy and/or sunny days; hopefully the atmosphere beneath will get steamy and humid, lowering the surface tension of the water

This is not adequate as a set-up-and-leave method of tending; hessian does not hold moisture well in the open air, and as it dries out it will promote the evaporation of moisture from the mortar. We have used this for large areas, but only because there was a three-hourly wetting system in place for the week following.

– which is all to the good. Keeping a wet material in contact with the mortar can be just as pointless, as the mortar needs to dry gently. Shades and windbreaks need to be erected as necessary to provide short-term protection.

Any method must be used and adapted in accordance with the conditions in the environs of the work and physically checked at the work zone. Remember that conditions at ground level can be very different from those a few scaffold lifts up.

Wire-brushing stone and joints is never to be recommended. Here the labourer is trying to remove cement stains to disguise the incorrect use of materials, among other bodge-ups committed in this incompetent attempt at repair work.

The lower wall here was badly repointed, and with no decent tending it has failed in the first frost. Note the pointing on the building showing the position of ties inserted across a vertical fracture; hopefully it was not done to the same standards as the garden wall.

If It Goes Wrong

Obviously mistakes can happen, and when working around a building or on a large area, it is possible to neglect the tending for some of the work. The masonry may have been too dry and sucked the moisture out of the mortar too quickly, or even that it was applied incorrectly – how did that happen? The result will be mortar that is doomed to premature failure. Evidence of this will be an overly pale colour, with a dry powdery crispness to the surface from drying too quickly. There may be shrinkage cracks in the mortar or it pulls away from the stone; there can be hollowness or movement when tapped; and, if left to get too cold, the surface will have a thin skin that falls off exposing crumbling material beneath. If failure of the mortar occurs, take it out and do it again properly; there is no other option.

Interesting treatment of mortar here on the chalk downs where brick and flint are common building materials. To repoint this glorious flint galleted wall would be challenging to say the least!

CHAPTER SEVEN

DUTCHMAN REPAIRS

For the physical repair of the stone in a structure, stone replacement and mortar repair are the options to choose between in order to conserve or restore its integrity; both have very wide application possibilities and methods. This chapter examines what is involved in stone replacement.

Stone Replacement

Whole Stones

Whole stones that cannot be considered structurally capable, having become decayed or damaged beyond repair, will need to be replaced as a unit. This is a cross-disciplinary approach between masonry and conservation, both having a say and a hand in the need for this. The performance of a stone will depend on its size and position, so massive Portland masonry blocks can lose substantial material and still support the wall, whilst a vented Hamstone corbel can be almost whole and still fail.

Facing Up

A slip of stone applied after removing the existing surface back to a suitable depth is used when only the surface is to be replaced with stone; this is effectively a masonry veneer or solid cladding.

OPPOSITE PAGE:
Test block and tool kit for piecing-in skills training in Canada using Tyndall stone; the edges of the piece are scribed then drawn in pencil.

Replacement tracery window in red sandstone requires all the stonemasonry skills, as well as good site practice. Note the keying on the joints to help the mortar grip and the packers for exact spacing. (Photo: Matt Harris)

Halving In

This is a process principally used for windows or tall thin elements, usually end-bedded, that decay in vertical layers, where the stone is cut back (usually to the glazing line) and the replacement pinned into the remaining stone.

Piecing In

Part replacement of decayed areas with a piece-in, indent or (more commonly) a Dutchman is one of the most frequently

Vandals had smashed off the nose and pieces of the face of William of Orange's statue at Brixham in Devon. The scars were prepared to take new stone pieces, then a squeeze mould was taken to make a cast of the features. New pieces were worked to fit the cast in the comfort of the workshop; they were then fixed with dowels and resin before being blended in.

used stone repair techniques, ranging from small plugs in dowel holes to multi-kilogram attachments to capitals. Facing up and small halve-ins can also be termed Dutchmen. As the methods for these options have a generic process, the main theme will be inserting a Dutchman, with notes to highlight variations as appropriate.

Lovely Dutchman to a capital of Purbeck marble, which is actually a polishable limestone.

Replacement stones to quoin. Here the joints must match those of the wall, and also it helps if the stone edges are worked the same way as the existing, here sawn.

MOULDING RULE

When replacing a complete stone it is general practice to recreate the original lines of the moulding, to promote the architectural design of the structure. It may be necessary to measure off another stone to get this information. For some instances where making a Dutchman, the repair can be cut down to the existing profile of the moulding to give the stone a balanced sense of wear. In all replacement scenarios there will be much discussion and possibly argument as to the extent of intervention, type of repair and finish.

Moulding to any stone replacement on Salisbury Cathedral is always to the original lines; if this were not practice, a building this old would eventually end up a jelly-like caricature as detail becomes lost.

Stone hood mould dressed back to determine the extent of replacement needed. This can be either face attached (almost halving) or the whole stone can be removed. The stone to the right is still barely holding its form, but with a building like this, the process is carried on continuously so its turn may be for the next time they get round to this area, possibly years away.

Box trammel used in stonemasonry is tougher and bigger than standard trammels; these are easy enough to get made or find a mason and ask where they are available.

Marking Up

The first step is to determine the extent of the decayed stone to be replaced; on ashlar it tends to be right up to sound stone, whereas some mouldings will be repaired to a limit, then butted up to with a mortar repair. It is essential that the Dutchman be attached to sound stone, usually at the back of the repair.

Ashlar

For a regular repair (with straight edges), lines should be parallel with the edges of the stone, or true vertical and horizontal. Measure two points equal from one edge and pencil in a line to the farthest point of the repair, then lay a flat metal square along this and scribe the two lines; use a level to pencil in the first line if measuring is difficult, or use a box trammel.

Moulding

Draw the first line by measuring or with a level along the best surface, fillet or similar. The other line is going to follow a curved surface so draw then scribe the next line with a strip of plastic paper following the contour at right angles to the base (first) line.

For more complex shapes, hold a flat board in the crook of a square that is resting on a flat surface, and adjust until a pencil flat to the board will touch the two corners concerned; then, keeping the pencil (or a scriber) flat on the board, follow the contours to draw a connecting line. This is a bit fiddly at first, but with practice gives excellent results.

Marking out a mitre while working a return is the same process as getting a straight line drawn on a curved surface.

Sensible and honest repair to a Portland gravestone; dowels alone would not be realistic, so rails were attached, with studs resined in holes in the back of the stone and the angle bar (even if this is stainless steel, it will be called angle-iron by most, so be aware) bolted on. The missing piece was made up as a blank with no decoration and dowelled into both pieces.

Curved Ashlar Repair

Some repairs, particularly to corners that have blown from expanding cramps, will benefit from a shaped repair; this reduces the amount of stone to cut out and gives a 'softer' appearance to the result. Determine the shape to be used, remembering not to bring the stone to a fine point (never make triangular repairs with sharp angles) as this makes for a weak stone that is difficult to make and fix. Then make a templet of the shape in plastic paper and scribe around it, marking it with the position, depth required and bed (if the stone is to be worked at the workshop). If there are a lot of repairs, use a numbering system, and mark the stone to be cut out with the same number as the templet.

Cutting Out

With a fine sharp chisel placed in the scribed line almost at right angles to the surface, pitch in towards the unwanted stone with a sharp strike around the perimeter; this gives a nice sharp edge and will prevent spalls during the next stage.

Here a sharp line is cut into a flute with almost a pitching strike.

Chop a relieving draft along the line using a 12mm chisel, remembering to cut into the stone at the ends that meet the joint/bed to prevent the corners blowing.

Working by Hand

For softer stones that are firmly set in place, or awkward mouldings, it is possible to remove the unwanted stone by hand. Working from the edges in, use an appropriate tool – this may be point, claw or chisel – and commence cutting out the stone to the required depth. Drill a set of holes in the stone to the required depth for quick checking that all

Cutting out a decayed stone requires good tool skills and care to prevent dislodging other stones; always work in to the centre.

Dressing the edge of the space with a TCT rubbing block. Note the clawed surface is just in from the flat surface that will make the joint.

is well. Do not work to the finished line until the end, when with a sliding square the new joints can be worked square to the surface. A flat draft is only needed for the first 10–20mm in; the rest can be taken down with a claw, and if this is worked lower (which is easier to do) this will aid in locking the Dutchman in place. The back of the repair can be left rough for keying, and it is not essential to get the internal mitres sharp and square to this; a curved corner will allow mortar to slide around the stone when fixing.

Using Power Tools

Drills and disc cutters are the best way to help in the removal of the stone, reducing the amount of impact the stone has to take, and speeding up the whole process.

Drill

Mark the depth on a 10mm (approximately) drill bit by wrapping masking tape around the shank and proceed to make closely spaced holes in the stone. Drill the corner first and then work along the edges in from the arris; next drill lines across the stone, enough to allow the material to be knocked out without jamming the chisel.

Cutting to the glazing line for halving in mullion repair.

Some stones can be drilled using high-speed steel (HSS) bits, sharpening as required, while others will need masonry (TCT) bits; keep them sharp at all times and if possible do not use the hammer function, as the vibration can fracture old stone or cause mortar to drop out and loosen the stone. Then work by hand as above.

Disc Cutter

With a small disc cutter (115/125mm blade) cut a slot along the lines; how close to cut to the line will be down to practice

Making the first cuts to relieve the stone; note that no cuts are on the joints, as the chance of error should be minimized.

The first level of stone has been knocked off and now deeper cutting can carry on. Note the diagonal cuts to the corners, helping to relieve more material.

and often the position. It is possible to get to the level of cutting straight lines crisply to the scribe line after a time – just do not attempt it straight away! Stop at the corner and joints of the stone.

Now cut a grid of slots into the stone to the depth required; usually the size of the blade will be sufficient. Discs that wear down will need changing often, so invest in a good-quality

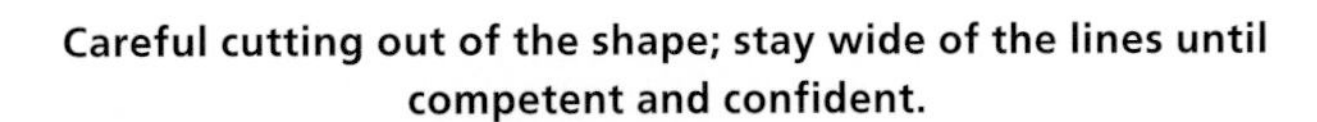

Careful cutting out of the shape; stay wide of the lines until competent and confident.

A handy home-designed tool, made by Brett for checking the precision of a cutout for a Dutchman; this has a half-millimetre tolerance, which is a good target to aim for. (Photo: Brett McBirnie, Ursus Stone Inc., Ottawa)

diamond blade as this will be more cost-effective over time and less hassle than a box of ever-diminishing blades. Knock off all the nubs of stone now with sharp strikes and, if this is the depth needed, work to the corners and back by hand. If more needs to come out, make another set of slots and then finish off. It is useful to have a block of hardwood with squared sides, slightly smaller than the repair, which can be offered into the space to check all is square and accurate.

Hood mould replacement with wood blocks supporting the upper stones until the new ones are fixed.

Mouldings are tackled in the same way, though it will be necessary to work in layers; cutting through projections and into recesses will require dexterity that relies on assessing the situation on site and using common sense.

Putting in the Stone

Making it fit is the first stage, and this progresses at different levels. Hopefully if the Dutchman is square, a block can be had to the exact dimensions of the hole. *Never* allow for a joint when measuring up new stone for the repair; the target is always to have as tight a joint as possible, preferably 1mm or less. For the inexperienced it is best to make the hole first, then any reworking due to mistakes or over-zealous cutting will not be a problem. Ensure the hole is completely square or regular as needed, then order or cut the Dutchman to that size. If the Dutchmen are coming from a quarry or works and there are more than two or so, have a duplicated cutting list with all reference numbers marked on the new stone, and check they are all square and (if the stone has well-defined bedding) the orientation is correct before signing them over.

Dressing in the edges of the space to get it perfect is essential, so take time and get nice drafts cut.

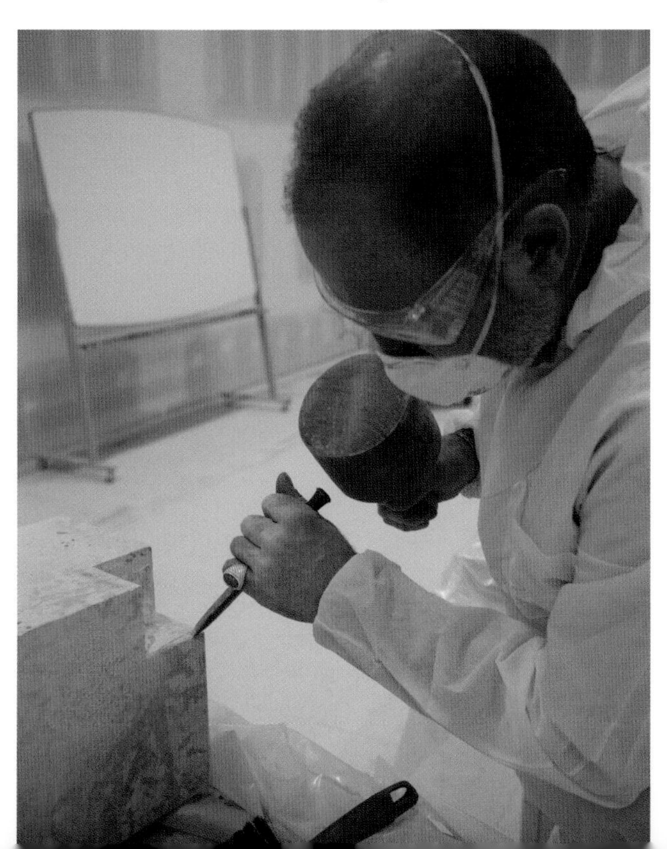

Depth

For a lot of square repairs it is practical to have a common depth and to cut all the Dutchmen from a single slab of that thickness. Single repairs may need to be cut from thicker stone, just to aid the working process. For working small stones, a sand-filled bowl or sand-filled inner tube lengths will be invaluable to hold them in place. Ad hoc vices can be cobbled together from blocks screwed to the banker and the stone clamped with quick-release clamps; just ensure the block cannot move when using power tools, or accidents can happen. Slice excess material off the back with a disc cutter and, using the grinding capabilities, round off all the edges on the back face and the return arrises to halfway; this is to allow the mortar to flow easily.

Processes of a nice efficient indent repair to marble carving. It shows tight work and how marble looks when carved before polishing by Jeroen, who, incidentally, being a real Dutchman, calls this technique *inboetwerk*. (Photos: Jeroen Kok)

Use plastic film when sliding the block in, enabling it to be pulled back out without damage. A recent government manual says to use paper for this – just don't! It will rip and literally leave you in a tight spot.

The flying Dutchman: a lovely dovetailed slot on a handrail in a French park, sadly missing its filler. This is the shape to cut in if the stone is in an exposed situation such as this (or the nose of a step, say), where the Dutchman can be dropped in. It should be for copings as well but is not used that often due to time and the (perhaps?) over-reliance on adhesives and pins.

Offering Up

Trying the stone for size should be carefully done, as any pinching can spall arrises when extracting jammed stone. As it is very difficult it is to cut a perfect hole, a tight fit will always have the chance of jamming somewhere; so a little help is needed. Pencil a line around the Dutchman 10–12mm back from the face, then with a grinder or *chemin de fer*, put the slightest of chamfers, sloping from this to the rear, on all joint faces.

Try the stone for size, but do not force it! If it looks to be going in completely, stop and put a strap (plastic paper or pallet strap – not paper!) around so it can be pulled out easily, without resorting to levering with a sharp tool and losing arrises. Usually, though, there will be pinch spots, so when it starts to jam, withdraw it and look at the joint surface; it will be dusty from the working and the pinch will have made a mark. See where the corresponding point is in the hole and decide whether to take a bit more off the Dutchman with the grinder, or knock down the high spot in the hole using a claw. Repeat as necessary until the stone slots in to form a snug fit and pulls out easily.

Cornice stone sized up and ready for dowel hole drilling.

Dowels

There are two choices here: permanent or removable. For a permanent dowel, a hole (or as many as are needed for support) is drilled in the back of the space and a protruding dowel fixed in, using resin or mortar. A corresponding hole in the back of the stone is filled with the resin or mortar and the stone pushed onto the dowel(s). When fixing the dowel into the back of the space, slide and rest the stone partly *in situ* to bring the dowel into line while the resin/mortar is setting; offer the stone in fully afterwards, to make sure there is no catching.

Removable and/or face dowels are useful if the stone may need replacing in the future, perhaps if the original stone is going to continue to decay, or if there is a need for angled (scissor) pinning, or to satisfy the reversibility clause in the conservation credo. Drill the Dutchman with a bit the same diameter as the dowel or fixing, working from the face towards the back; put gaffer tape over the region of the drill's exit to reduce the spall. If a screw or bolt fixing is to be used then put in a counterbore, deep enough to cover the head and allow a plug or mortar fill, but not so deep as to weaken the fixed strength. Place the stone in dry and drill through the hole into the back space walls, making sure this is deep enough for a guide for when the stone is removed and the hole in the space is taken out to the required size and depth.

For a screw-out fixing, set a plastic expanding rawlplug into the hole, fix the stone into position (as below) and then insert the screw, with an O-ring under the head to take up the irregularities (or a small dob of silicone mastic). Cap off with a small mortar repair or from a spare piece of the same stone. Use a core drill and get some plugs (these should be the same diameter as the counterbore) about 10–15mm long and gently drive them in with a sloppy mortar; when set they can be gently worked flush with a chisel.

Dry placing of a stringcourse stone to check alignment. Note the dowel hole drilled in the wall.

Here the mullion is being halved in and the dowels are set in the original, so the stone can be slid onto them.

Cleaning Out Holes

Holes need to completely free of dust to ensure the resin or mastic adheres to the stone; dust will soak up the resin and create a plug that can slide out of the hole too easily. Compressed air is the best to use; the little puffers that are often used are generally ineffectual and should be avoided. Washing out with water squirted from a syringe (buy the big plastic ones from farm suppliers) is an option, though this can create a plug of slurry at the bottom of downward-sloping holes. Mix the water with a small amount of alcohol (ethanol) to break down the surface tension of the water when it encounters fine dust. Bottle- or pipe-cleaning brushes are effective, especially on harder stone.

Fixing the Stone

The stone should be thoroughly damp, best achieved by leaving it in a bucket of water for a day, then set aside covered so that it loses excess water but does not dry out. The space and surface around should be dust free and sprayed successively until well moistened but without sitting water. Mask around the space with wide tape and fix a plastic apron below to prevent lime dropping onto the face. Pug out the back with a smooth creamy mortar which is mobile enough to squeeze through the joint, lightly spray the joint faces of the Dutchman and the face around the space.

Put any resin needed into dowel holes quickly and efficiently; using a resin that is not affected by moisture is important

DROP AND PULL DOWELS

On new stones or Dutchmen, where the size or shape could mean the stone may fall out or move laterally (jambs, mullions or proportionately thin ashlar blocks), it can be necessary to put dowels through the bedding joint. Obviously a stone with a dowel projecting from its side cannot be slid into a gap, so other techniques have been developed.

Make a simple templet (cardboard will do) for the top of the stone that fits flush with the face. Mark on this where the dowel will go and drill the hole, half the length of the dowel, into the top of the stone; give this a funnel shape at the top to help the dowel to locate. Drill a corresponding hole into the top of the space, just longer than the dowel. When the stone is ready to go in, attach fishing line to one end of the dowel allowing enough to come out of the space and be pulled. Fill the top hole with smooth creamy mortar and push the dowel in, tied end first. Fill the bottom hole with the same mortar and place the stone in the hole (carrying out all other mortar filling as well). Then pull the line and the dowel should slide down into the bottom hole; the displaced mortar mingling with that in the top hole will create the fix. For a bottom bed dowel, drill the long hole in the stone. A variation is to dry place it, doing the same as above without the mortar or line, keeping the dowel in place with a slip of plastic paper or long strip of gaffer tape that is pulled out once the stone is in place. Obviously, this will need grouting afterwards.

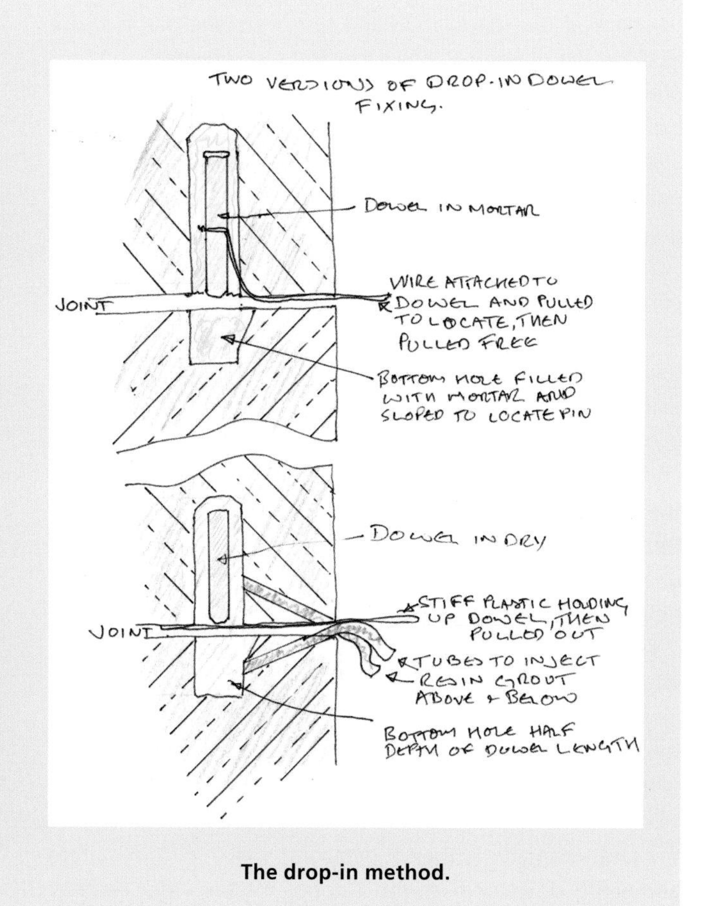

The drop-in method.

Mortar has been placed in the space in such a way as to squeeze it around the stone when it is pushed in.

here, so unless you can juggle all this adeptly, do not use the standard polyester resin commonly sold as stone glue – it should not be called 'glue', the correct term is 'adhesive'.

Apply a slurry coat then a smear of mortar to the sides of the Dutchman to just back from the face. Immediately put the stone in and push it home; if it is done correctly there will be an increase in resistance as it nears its position and it will need to be driven in by holding a piece of softwood against the face and beating it with a hammer or mallet to get it fully in. The mortar should squeeze out of the joints like a curled pat of butter and start crisping up immediately; strike this off with a sharp trowel and, with a clean damp sponge to the joint, lift off any mortar that could stain. Keep the sponge scrupulously clean by repeatedly dunking it in clean water and squeezing to rinse out the mortar. Point up any empty joint with a stiffer version of the mortar, once again keeping the face clean by sponging off.

RESIN EMERGENCIES

If for some reason resin happens to squirt out of a joint or drops onto stone – do not wipe it off! Though it is a reflex action to get it off as soon as possible, the resin will smear into the pores and leave an indelible stain; let it start into its cure and as it changes from liquid to solid (glass transition phase or GTP), peel it off with a sharp instrument. If there is a stain, now is the time to take a bit of absorbent paper and dab it off with a touch of solvent such as acetone; do not go wild and wet everything; keep it neat as it may be the resin has cured and will not be removed like this. I would caution against sanding off a stain like this, but it will be purely down to a site decision.

The messy result of allowing incompetents to fix stone; there is resin all over the place, with no room for a mortar joint, and the only way to clean it off is to remove the stone surface it is smeared on.

Nasty cementitious repair to a ledger slab; cement usually has two states on stone, either it is so strongly adhered that it is impossible to get off without damage, or, as here, completely detached and useless.

A dentil repair has been pushed into place by firm strikes on a wooden block and the mortar has come out evenly from the joints.

It is essential to sponge the lime off the stone carefully; always use clean water and sponge.

Dry Placing and Grout

Some stones will be too big or awkward to place using the above method, so they have to be put into position with joints spaced using packers; these can be lead or slate but avoid trying to match the gap needed with a pile of slate bits. Always have a bumper pack of plastic packers (used in building construction generally) that come in a variety of thicknesses. The grout that is to be poured in will obviously

Large blocks going in on thick beds and being packed around the back; the wooden wedges here provide temporary support until the pointing is placed.

pour out again, so the joints need to be sealed. They can be pointed up, but the mortar must cure before grouting, entailing a wait of at least a day or it will be forced out by the grout. The quick method is to push compressible draft excluder strip in with the edge of a trowel around the bottom and sides.

Drill a hole of the same diameter as the grout pipe at low level in the side joint(s), flush out dust with a squirt of water, set a tube in sealed with plasticine (do not use clay as it will wash out when wetted by the grout) and, with the cup at a higher level, pour the grout in. A length of wire can be put down the tube, and agitated occasionally to keep the grout flowing. Do not stop for any length of time and then try to start again in the same hole as the grout will dry at the end and stop any more coming in; on big stones have a set of ascending holes and fill them from the bottom up, plugging them as you go by pushing a golf tee into the end of the tube.

Finishing Off

Hopefully the Dutchman is perfectly aligned with the face, but often it may sit proud, so it will need to be worked back; do this when the mortar has cured soundly. If it looks safe drag as much of the surface back with whatever will do the job; do not let the drag scrape the surface of the original stone. The last bit will need to be hand tooled with a very sharp boasting chisel and light mallet or dummy.

Matching Moulding

Working a full mould *in situ*, to match up to an existing, will probably cause the newly fixed stone to become dislodged, so do as much work as possible before fixing. Cut out the space and get a squared block to fit as above, then on the

These new pinnacles to this church in Chittlehampton were to replace storm-damaged ones that had been blown off. The structural engineer working for the insurance company ignored the fact that the Victorian pinnacles, correctly fixed and six metres tall, had withstood the onslaught, emphasizing that it was poor modern workmanship that had caused the last replacements to topple – mainly because dowels were inadequate or the holes had not been cleaned out. A continuous steel armature within the structure was recommended from the cap, down through the legs and two metres into the church tower. This involved lots of interesting steel design, and we used a low-viscosity resin grout to fill the holes and set the rod. The upside is that they are still in place, but soon after fixing it was noticed that, due to their inflexibility (stone structures need to absorb wind and stress), there were hairline fractures developing. All this was watched over by the survivors fixed simply yet effectively over a century ago.

Using a rasp to work the high spots down in the curve of a Bathstone moulding. This is an excellent tool for dressing in Dutchmen to match on the softer stones; these files are available in different profiles and sizes, and can often be found in second-hand tool shops at a good price.

Dressing off the top of a Dutchman to a capital requires good stonemasonry skills, as well as ensuring the piece is fixed securely; work in from the edges to prevent spalling and loosening.

Dutchman trace around the section needed and pull out to work it. This could be a small stone that will be difficult to keep still while working the shape, even in a sand bowl. If this is the case, stick the back of it securely to a larger block using polyester resin and carry out the work, then cut it off with a disc cutter or, if the stone can take it, by driving a wedge into the adhesive joint to split it off. Chisel off the remains of the resin on the back and fix as before.

New merlons, securely fixed by dowels and cramps. It was decided that these were to be in Portland whereas the majority of the stone here was Beer – difficult to procure and not really up to the job of weathering in exposed positions.

New halved-in mullions that have been fixed with natural bedding using short stones, rather than the other option which is to use as long a stone as possible face- or edge-bedded.

Saw-Lash

Always get rid of saw-lash on the face of stones (in every situation, not just this process) by rubbing with carborundum block, diamond or cintride pad, before the stones are used. The undulations of saw-lash can mar the flatness of the face in a way that is noticeable only when fixed and cleaned.

A nice Dutchman marred by the saw-lash left on it; this is at a low level so is quite noticeable. Also I am not too keen on the rounded arris, as others in this work have been left sharp. Consider if all repair work is rounded to match erosion – how long before the building turns into a jelly mould?

Replacement stones at Woodchester Mansion have been given an agreeable tooled finish to start them off well. Note the original blocks have hand-worked bed and joint surfaces, adding a touch of character, while the new is sawn stone; these flat lines will continue to show as the stone erodes, always marking them out as different from the livelier-edged originals.

Texturing

Fresh stone in a wall will always stand out; without any texture it will stay this way for a long time and weather differently from the rest. It is best to give the new stone some modicum of texture, mainly because bland featureless surfaces do not look good in a building, but also to 'break it in' better. The texture will depend on what is decided on site, or the existing surface condition.

Boasting evenly set-out toolmarks across the surface is one way that masons would finish off stonemasonry in the past, and this is easy to replicate, as the tools are the same today as they have always been.

Producing an axed finish is another way and this can be approximated by loosely holding a chisel or boaster and working backwards across the stone, letting the tool jump when struck; practise on a spare – it is surprising how effective this can be.

A finish popular in the nineteenth and twentieth centuries on softer stones was dragging with a Bathstone drag or cockscomb. This enhances mouldings beautifully with its fine parallel lines giving a building much-needed shading and interplay of light. To replicate this it is best to use drags of the same size, although cheaper, practical alternatives are sections of saw blades (hacksaw, tenon, etc.) glued into wooden handles; it is possible to get the spacing correct as saw blades are sold with various numbers of teeth per inch.

Ashlar was often fairly smooth, only gaining texture as the elements gently pockmarked the face, so the solution here is to give the surface of the new stone a going-over with air abrasive that, depending on the grit and number of runs, can effect a gently (or stronger) weathered surface akin to the original.

Venetian wellhead of Istrian stone has been neatly repaired with tight joints and good texturing. If you get the chance in Italy search out any examples of *pietra dura*, or visit the museum in Florence to see the true beauty of tight joints and craftsmanship.

New stones to cap buttresses have been sparrow-pecked, using a point, to give them a start in joining the soft weathered appearance of the original work.

ACCIDENTS WILL HAPPEN

Working on the King's Statue in Weymouth, the Coadestone shield had been taken off, cleaned and had its painting and gilding finished. It was lifted into place and propped up with wooden battens before being fixed. That night there was an unforeseen storm and it was blown over (never underestimate the power of the weather!) and broke into many pieces.

Not a good way to start the day; the shield lies in pieces.

This was a horrifying sight to be greeted with in the morning, especially as there was an inspection of the site due in a couple of days, so once the panic was over, a quick repair session was convened. (It had actually happened on 1 April and convincing people over the phone took some doing.)

The shield was assembled on clay formers to get the right shape and match everything up, then straight lines were marked across the joints to show the dowel locations. These were drilled with connecting holes exiting the back of the shield. It was then assembled dry, face down, with the dowels in place and strapped around the perimeter.

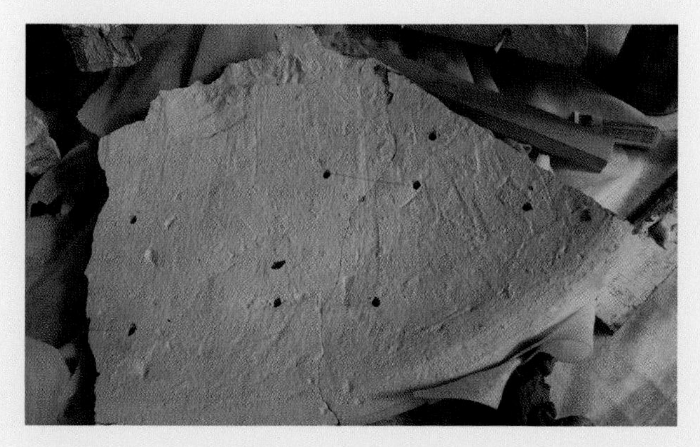

Fill and exit holes for injected resin; dry-set dowels are in the piece at the top.

Resin was injected under pressure into the holes which filled the dowel hole and exited at the other end. Once cured any fissures were filled with an epoxy putty (no need for porosity in stoneware), cleaned up and refinished, the whole achieved with a big sigh of relief and in time for the inspection.

After the resin had cured, there were inevitably some areas where it had come through to the face; these were brittle and broke off easily.

All pieced together and missing fragments made up in epoxy modelling paste, the shield is fixed in position.

Finished and all is well with the world again.

CASE STUDY: CRAMPS

Used as a no-nonsense way of tying stones together, and providing restraint to copings and overbalanced elements, the durability of cramps is essential; unfortunately many have been made of (ferrous) metal, which can corrode and rot away if it comes into contact with water. One of the regular tasks in repair of stonemasonry is the replacement of these with non-ferrous (usually stainless steel, but occasionally bronze) alternatives. Those on the top of stones are easy enough to replace, whilst those buried in the fabric are left alone unless there is evidence, such as fracturing of the stone, that they are in the process of decay. As oxidizing ferrous metal can inexorably expand to twelve times its original volume, it can exert tremendous pressure on the stone encapsulating it.

Typical rusting cramp, its ends set in lead but the rest exposed to the elements.

Levering the old cramp out. Some will need cutting up, if they have not degraded already. If the lead will not come out or holds the downturn in, then it may need to be drilled out using HSS drill bits.

Reworking the channel to take a new cramp.

A new cramp made out of flat bar and bent at the ends – known as twice-turned down – is pushed into a resin-filled channel. These can be left bare-backed or, if appropriate, set lower in and covered with mortar, which will not last due to it being directly onto metal and poorly attached to the stone; a better option would be lead.

A strip of lead is melted *in situ* with a hand-held plumber's torch. (Incidentally, the term 'plumber' comes from the Latin *plumbum*, lead.)

It will shrink a bit so it will be necessary to spread it in as it cools.

Bronze cramps set in lead, on one of the lovely fountains that are a feature of Aix-en-Provence, are tarnishing nicely and have coped this way for centuries.

Natural Dutchmen? The cramps on this plinth started blowing out the corners of stone. Rather than cut the stone out for a Dutchman repair, the fragments were carefully taken off and numbered. The rusting metal was dug, drilled and chiselled out without enlarging the limits of the fracture scar and new cramps were inserted (almost keyhole masonry!). The remaining metal had a reducing treatment applied to halt the oxidation process. The fragments were then placed in position and drilled through the face into the fixed block. Rawlplugs were inserted into the block, the outer holes were counter-bored and the fragments were attached with a screw with the joints lightly mortared up; the heads were then covered with mortar plugs. This is reversible and allows future inspection without damage – good conservation practice.

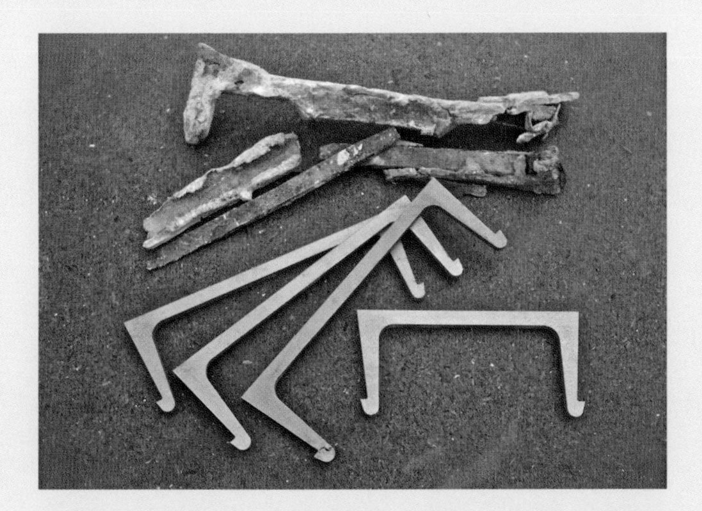

A personal design for replacement or new cramps, cut from stainless steel. They are rigid, of regular sizes and easy to install with minimum damage and tools. The debris at the top is the remains of ferrous cramps that, even though they were set in lead, have still succumbed to the ravages of chemistry and physics.

To set these cramps, a templet is placed on the stone, then the position and shape marked with a scriber. Holes are drilled for the ends, and by swivelling the drill the shape is made for the turn-downs. Next a channel is cut with a disc cutter, the dust is blown out and with no banging or shock to the stone it is ready.

Dropped in, the cramp sits perfectly in position; it can be pushed into resin or a liquid grout.

Copings anchored with the cramps.

CHAPTER EIGHT

MORTAR REPAIRS

Stone as a construction material, while not impervious to the weathering effects of time and nature, is considered by many to require little care to protect it (we all know building maintenance is not high on most owners' to-do lists). Unyielding resistance could almost be true of some of the highly durable igneous rocks, but there is a battery of mechanisms working to cause the downfall of masonry and its elements. As deterioration can be gradual, concealed or unnoticeable in superficial examination, it is easy for the surface of a memorial or wall to be neglected and reach a stage where some degree of intrusive repair becomes essential. This situation is not always an invitation for removal of original material showing signs of wear. Structural integrity of masonry can remain sound while the appearance or finish is compromised; remember that stones that show the history and theme of the original building are of paramount importance, so think minimum intervention – think mortar repair.

A north-facing church window in Devon crumbles away as the salty air lays waste to the stone tracery.

Standards

Mortar repair (sometimes termed plastic repair) is used to bring (mainly dressed) stone back up to its original lines, building up the missing area with a suitable mixture of lime and aggregate. We can term this the 'good' approach when done satisfactorily.

Termed a minimal intervention when used wisely, mortar repair is now seen as a profitable market for material suppliers

Another window in Devon has had sympathetic repairs of lime mortars carried out fifteen years ago and still holds up against the elements.

OPPOSITE PAGE:
Missing stonework at Wardour Castle is honestly shown with a recessed mortar fill. Note the damp conditions indicated by the algae.

Hard cementitious repairs to a chimneystack in Oxford present an unyielding mismatched patchwork as the original stone erodes away.

In Canada a sandstone plinth has been repaired with a mortar that not only is wrong, but actually hastening decay.

who purport to provide universal ready-made solutions. They will harp on about it being mineral and easy to use, but sadly most proprietary brands of plastic repair mixes available are based on an OPC binder; easy to apply and having a hard durable finish, they should not be considered for conservation repair applications, due to their incompatibility. Equally suspicious consideration should be given to any material using organic resins as the binder. We risk straying into the area of 'bad'.

Any repair can be detrimental to the welfare of the stone when produced badly or with unsuitable materials, actually hastening the decay in some cases. The prime example of this is hard cementitious mortar, used in repair patches and pointing, as the mortar starts to stand proud of the face while the stone is being destroyed around the edges of the repair by concentration of salts, forced to exit through the stone by the imperviousness of the mortar. Made with OPC or white cement, repairs will always stand out, usually getting worse over time – definitely ugly.

Does Colour Matter?

For many people the most obvious requirement for mortar repair, and whenever a specification is produced for the same, is that it be a good colour match to the stone; this is a good start but it misses the most obvious requirement of a repair that needs to work on stone exposed to the elements. The colour of the average mortar repair will start to differ from the original quite rapidly, unless specifically manufactured for *that* stone – not that wall, just that stone: as we have already noted, 'stone is a natural material and may vary in colour and quality', so no two pieces will be exactly the same.

The colour of a pre-packaged or large ready-mixed batch mortar cannot be expected to blend in over any one area. Indeed, sometimes this may not be a bad thing, as the tonal varieties of masonry façades can absorb another hue or shade of the base colour, as long as the *texture* is correct.

Texture is the Crucial Element

The texture of a mortar is the most crucial external aspect to be considered when used to repair stones, *assuming the colour is within the tonal range of the cleaned stone*. The visual aesthetics are dependent upon the aggregates used and the manner of application.

Consider how the only time that the surface of the stone is actually seen is in the years immediately following construction, or after a cleaning programme (though any cleaning that brings a historic building back to its original level of cleanliness is almost certainly too strong). The colours of the stone are the results of pollution, dirt or organic growth collecting on the texture of the surface. The repair colour should be based upon a not-too-clean example of the stone itself – not the surface accretions. If the building is dirty and unlikely to be cleaned, do not 'antique' the repairs; the staining will continue apace and, without exception, artificial staining will not keep up, preferring to go its own way!

Nicely textured mortar blends in well with the original stone on this Marnhull stone jamb.

Successful mortar repairs primarily replicate the durability and texture of the original stone (colour often being an afterthought) that works because the repair surface is for all intents and purposes the same as the stone when it comes to collecting debris from soiling and weathering processes – dirt does not react differently to colour whereas a change in texture will result in an area weathering in a different manner completely.

TRIALS AND SAMPLES

Physical samples of the cured mortar, rather than just descriptions, are essential as standards to work to; in larger contracts the production of these should be carried out as a separate contract. The conservator designs and presents finished, cured examples with full documentation on the materials, methods and tools used to get the appearance. This should be done at least a month before the work starts.

Well-made mortar samples will keep well and provide invaluable reference resources; do it properly and be professional.

Keep each sample an appropriate size, allocate a reference number and place them in the correct environment to cure for at least a fortnight. Store them securely as the similarity of the repairs needs to be checked prior to signing off the work; they are also a reference piece for the operatives to work to. There is nothing more depressingly amateurish than to see various sized pats of mortar sitting on the windowsill of a site hut like some failed primary school baking lesson, so do it professionally and think about presentation. Aim for a working palette that will cover all areas of the façade, not just with a variety of colours but of textures and coarseness as well – even good-quality Portland will have variation in pore size and texture.

If purchasing ready-made repair mortars from a manufacturer, the full recipe and quantities should be given as well; this is pragmatic as work may need to be carried out in the future using the same mix, or possibly the materials may react with future cleaning, etc. The full description of all materials used is a legal requirement for work carried out on listed buildings – 'trade secrets' is not a phrase that should ever be encountered in work on the built heritage.

This is how not to do it! The samples are inadequate, do not appear to have been cured correctly and will end up as trash when someone knocks them off or tidies up.

The Purpose of Mortar

The role of the repair is to fill in a missing area of the stone, protecting and recreating visual harmony, without losing any more of the original than necessary.

Another important consideration is the prospect of regular inspection and maintenance; if a repair is to be exposed or unreachable by normal inspection, it can decay and exacerbate problems before it is noticed, so stone replacement should be considered too. The true purpose of mortar repair, though, is to be sacrificial, being as porous as and softer than the original stone; the mortar will break down faster, diverting nasty effects from the original stone.

The Nature of Mortar

As the material to be replaced will define the mortar it would seem logical to use the same materials that made up the original stone; however, this idea does not have many followers in the field. An aggregate made from crushed stone of the type to be repaired and a sand mix is the best base to start from, though some sands alone will give the appearance and texture of weathered stone if made correctly. It is best to base the mix on easily obtainable materials that do not require overly complicated recipes for the finished result, and the colour should preferably derive from aggregates that remain tonally stable rather than relying on colouring agents that could react or fade in use. Obviously some applications will require pigments, so understand and allow for any chemical alteration of the ingredients. It is prudent to have a mix that will not be adversely affected by crude preparation of materials, as slight variations can be used to good effect.

Red conglomerate repair at ground level in Exeter. After fourteen years the tone and texture is holding up well.

The Process of Mortar Repair

Always remember your key word here – PAST. This stands for Prepare, Apply, Shape and Tend.

This Coadestone lion was found to have had crude cement fill to a large missing area when the paint was cleaned off to prepare it for gilding. This was removed by drilling in to relieve the stress, then carefully chiselling back to sound material.

The mortar was designed to have no shrinkage and be extremely tough. Built up in one hit, it was allowed to get a good cure by meticulous tending.

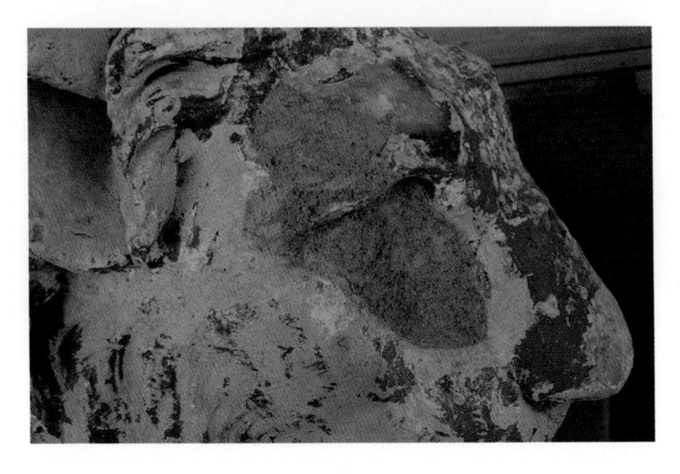

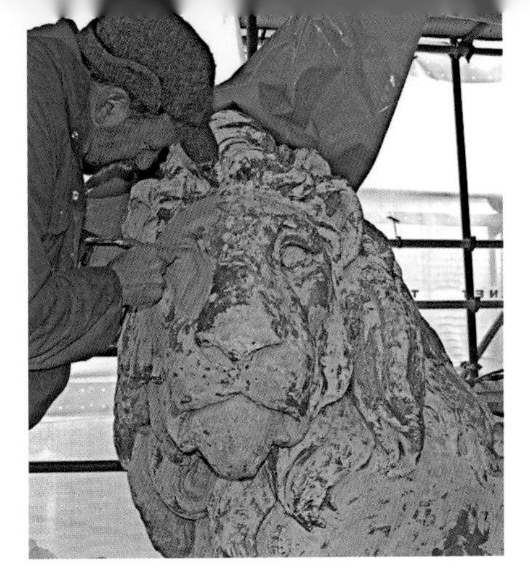

Modelling up the missing area to blend in and have the same texture as the Coadestone.

Identify

With a piece of chalk, walk around the job and mark where all the repairs are going to be; if needs be, get the extent confirmed by those in charge. Look at the scar and figure out how to get a nominal thickness of mortar running around the edge at right angles to the worked surface; with a mason's pencil mark this border. If it crumbles then cut out more – but remember least is best! It is not advisable to make a repair direct to the stone surface without preparation that ranges from cutting back right up to the use of full armatures.

Finished repair cured hard and primed ready for gilding.

The Edge

Use a fine bladed, fire-sharpened or TCT chisel to chop or chase in a regular-depth perpendicular fillet back to sound stone around the perimeter of the scar. With a fine claw or chisel, depending on the stone, work back the decaying stone in the scar to sound material, or at least to the depth of the fillet. Undercutting the fillet edge a few degrees from normal will lock the mortar under the stone. Be careful not to spall the arris when doing this. On soft stone a scraper can accomplish this – but do not get too carried away as stone will be weakened if brought to too fine a point.

Screws and armature wire set into the scar ready for slurrying and filling.

The repaired Coadestone lion fully restored to dazzling glory, and ready for a long life. The quality of the King's Statue restoration won several awards.

Keying In

Without an armature, the mortar will have to be keyed into the stone. The claw marks will achieve this to a degree, but to be safe, with a drill and bit of about 10mm make holes over the surface to about 10–15mm deep. Move the drill slightly

A rather half-hearted attempt at keying in for a mortar repair. The drill holes should be deeper, the edge cut sharper and possibly the whole scar taken deeper; the reason this failure is visible is because the mortar has fallen off.

Clay pegs set in calcium caseinate mastic to provide support for vertical mortar application.

from side to side to enlarge the bottom of the holes, forming dove-tails. Chopping across the stone with a chisel will give small depressions, similar to how walling stone was prepared for rendering; if you come across stonemasonry with gouges or little triangular cuts across, then it was originally rendered.

Supports and Armatures

Everybody needs support now and then, and none more so than the lowly mortar repair. Vertical or overhanging repairs are subject to the desire to return to mother earth, so inserting a restraint is necessary; practically, it is not recommended to make a repair that does not have some mechanical aid within the mass.

Projecting clay pegs made from broken tiles can be set into holes with mortar, resin (if you must) or preferably a calcium caseinate mastic (mix 9 parts sloppy lime putty with 1 part casein into a paste). Blow the dust out, wet the hole up and fill with the mix. Wet the peg, dunk it into the mixing tub and gently tap it into the hole, letting the mastic squeeze out; this will cure into a decent adhesive bond. Older porous, moisture-retentive tiles are best for this, and as they are low-fired, the area around them may dust off which can give a bit of hydraulicity.

MAKING CLAY ARMATURES

It is possible to buy (rather expensive) ready-made clay armatures, but if there is a need for a large number consider making them in-house; they will be useful as stock.

For this you will need the use of a kiln; try local colleges or potters and ask if you can include the pegs in their next firing. With a jigsaw, cut out the shapes of the pegs in 6mm plywood, sand down the templet and varnish it well. Place on a smooth board, give a light spray of water and firmly press well-kneaded brick-clay into the cutouts. Drag the excess clay off with a clean board or similar, lift off the templet and remove the pegs to a cool dry place to leather up. Repeat as necessary. When they are ready stack them neatly, with gaps between, into the kiln and fire them at about 850° – clay fired at a lower temperature can swell with moisture and crack the mortar.

Steel

A very effective way is to anchor stainless steel wire or strips of expanded metal/mesh across the repair.

Stainless steel screws can be set in the scar by drilling holes and inserting plastic plugs before screwing them in. Do not over-tighten as the expansion can pop the stone; very little

An armature has been added to reinforce the mortar and the slurry coat applied.

Shelter coat has been applied to blend it in with the surrounding render – a good-quality result.

holding does the trick. Now the screw heads will form dovetail pegs that can be used on their own for small repairs; ensure the top of the head is lower from the surface than twice the depth of the aggregate particles, so they do not impede the application of the final coat. For larger areas lace wire around the screws to form a light lattice. On deeper areas fine threaded bar can be bent to shape and the ends tapped into the plastic plugs.

A (flexible) rule is the jumping of joints, but it is good practice to make armatures and mortar repairs to individual stones; this is up to the operative and common sense. Alternatively, use calcium caseinate to anchor screws or bar in place, although it is tricky getting the mastic into the small holes effectively. Using polyester resin to glue bent ends of wire into holes is common, but has a high fiddle (and failure) rate so is best avoided.

THERMAL EXPANSION – A MYTH

One of the arguments for using clay armatures rather than metal is that the coefficient of expansion of materials used together should be similar; and so people surmised that because clay seems 'stonier' than steel it is probably the best to use. In truth the expansion of mortar is closer to that of steel than that of fired clay.

Filling Up

Blow all the dust out and ensure that the stone is suitably moist (see previous chapter): put a scoop of the mortar in a bowl and stir in water to make a brushable slurry. Keep stirring so it remains in suspension and work a coating of this onto the surface of the scar, getting right up under the edge. Sponge smears off the surrounding stone. Keep this very lightly mist-sprayed, because if it dries too fast it will become powdery and prevent the next coat adhering properly.

A stone urn has the base coat built up over an armature and this is allowed to cure before the final coat.

A notebook diagram showing repair of the shattered urn.

After making repairs to a Hamstone urn, to protect the stone from salts in the gardeners' fertilizer, the urn was lined with a lime mortar, then a GRP liner was laid over with an exit spout sealed in; here the edge of the liner can just be seen.

Strong Base

When applying mortar repair to large areas or depths, a mortar that is sufficiently plastic to use in replication of detail and texture may not have the cohesion and strength to ensure longevity; so a coarse strong base and softer finish coat is often used. The base may shrink slightly, but this is acceptable if confined here and allows the final coat to cure without too much shrinkage.

Deep voids should be filled with a mix that should be much coarser than the final coat, adding moist shards of stone into the body almost up to the finished level. Let this get well into a cure, allowing shrinkage to take place away from the final coat. Scrape back this base coat to give a consistent depth across the repair of no more than 15mm deep and, by roughing the surface, provide a key for the next coat. Give this a light slurry coat as before, and tend.

Keeping Up the Suspension

It is often the case that a liquid version of a mortar – slurry, grout, sheltercoat – is needed, so water is added and the mix then used; this must be kept in motion to prevent the base material settling at the bottom. Once the liquid has been used and the heavier residue at the bottom is reached, scrape it out and throw away – all the lime would have gone into the solution and the leftover has no binder in it. To make more, use a new scoop of mortar and start again – never re-dilute a slurry, grout or shelter coat!

The Topcoat

Now with a shallow scar or prepared fill, knock up enough repair mortar and set to: with a trowel apply a layer across the surface, pushing and working well into the background. With a brush dampened in slurry of this mix, brush the edges in to prevent the mortar pulling away while working, then get more mortar in to bring up the level proud of the stone around. Be careful at the edges with regard to staining, but ensure that the mix is in full contact at the perimeter; this may require careful buttering and pushing in at the edges to accomplish.

Hamstone window with well-made mortar repairs.

The mortar has been built up and once into a set it will be textured.

Getting Edges

Arrises and returns or soffits need to have a former to work against; use clean polished pine or extruded aluminium section. Get the former into position, if possible wedging it in place. Extending plasterer's supports can be lodged from something solid to do this – do not wedge off the scaffold planks as these can move as you walk on them; go off scaffold tubes. Tend this well, in the same manner as pointing, until it has become quite firm but not dry – practice and experience will soon show the perfect moment.

Rubbing Back

For a flat repair, use a smooth planed pine rubber longer than the scar (as time goes by you will accumulate favourite rubbers for all occasions, so in workshop time clean them up and, if appropriate, attach handles or grips to facilitate their use). Cut, do not drag, the excess mortar off by sliding across from the edge, pressing hard to compact the surface, in a side-to-side guillotine motion, preferably in one pass.

Mouldings and Returns

Any flats should be cut in as above, keeping back from projecting mouldings, by using the existing stone as a level. Lay

The mortar is built up and will be left to cure under the protection.

LOCALIZED CURING

Cotton wool is occasionally used by some to protect the cure; please don't! It will suck out moisture or, if it manages to stay wet, it will give the mortar a nasty plastic sheen; for localized small patches it is best to make a small 'greenhouse'.

Frame the repair with gaffer tape, cut a patch of plastic sheet to be just smaller than this, soak a sanitary towel (these hold a good amount of moisture without letting it out) and place over the repair then tape the sheet around all edges onto the gaffer frame; this makes a maintenance-free curing solution that can be left for as long as you are on site, hopefully steaming up, giving a really good cure.

The mini 'greenhouses' offer the best way to tend repairs and give a perfect curing environment.

a metal straight edge along the arris lines and, using the fine blade on a plasterer's tool, trim the arris in, then texture up with a rubber. Concave surfaces are dragged out with a fine cockscomb stone drag and then a fine-edged rubber; a quick alternative is to bend a flexible mitre-saw blade to the correct profile, clamp it to wood, then drag with it.

Cutting mitres in will involve some dextrous scraping and dragging; to get the exact line a flat piece of wood can be held and a cockscomb slid along it to define the line to work to. Tend until cured.

Sheltercoats and Washes

For homogeneous appearances (that is, one making everything look the same) it is often the practice to wash over the new and the old with an appropriate sheltercoat or wash. Tonal colours and shading can be applied using appropriate binders and mediums, the design and application of which should always be the result of trial patches, prepared well in advance, so the final colour is achieved.

Owing to the layman's ignorance of the hard work that goes into bringing stonework back to glory, this final, almost superfluous touching up is often the result that the quality of the whole work is judged on.

In Bury St Edmunds, this pediment sculpture had lost much detail and was restored with mortar repair; this was sheltercoated (the extent of which can be seen on the stones around the carved work).

Building up a massive mortar repair of a swag in Oxford; this needed good control of the mortar to prevent shrinkage by using pozzolans and aggregate.

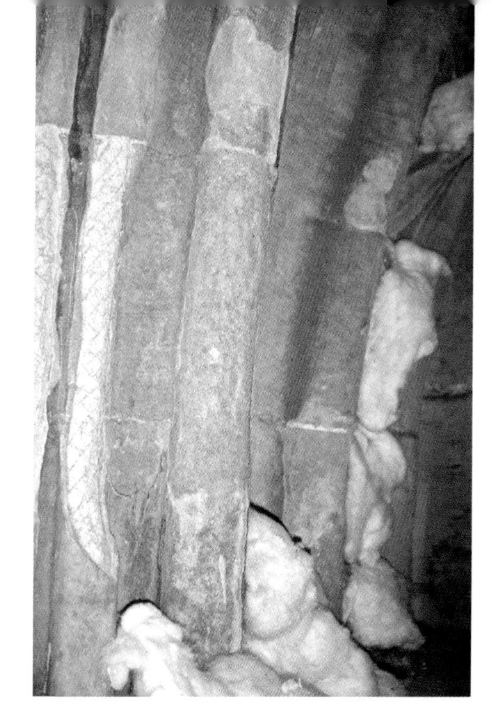

Mortar repair in progress. While this is a much-used style, there are a few issues that need to be considered: the keyed-in surfaces to the left are too low and appear to have dried out too quickly – they will need to be redone. The finished repair is textured to present an eroded stone – an effect that can only be subjective and if used to excess the detail of the architecture is lost. The lines of the original bowtell (curved) moulding are visible at either end of the repair, and could have been respected by working to them. The cotton wool for curing protection is not practical or efficient enough to do the job properly – as can be seen to the left.

Key Points

- Do not wet-sponge over a mortar repair; this will cause an ugly and ill-matched surface finish.
- Work must be covered *at all times* to prevent failure caused by drying out – this should not consist of hessian alone as this can accelerate the process. Damp hessian or similar with a polythene cover creates a humid environment, and it should remain in place for a minimum of three days after the last application. Mortar mixing on site must be done with clean equipment and standardized measurement of volume or weight; sand should be checked for bulking, lime putty should be old and hydraulic lime young.
- Always have a mist sprayer, bucket of clean water and sponge to hand as well as the designated finishing tool.
- Keep a check on the standard and quality of work at all times – if it is too white, powdery, smooth or ugly, then take it out and start again!

Mortar repairs that never looked like stone to start off with will not blend in to the building. Here they are overly smoothed and have an odd hue; not good work.

This medieval window in Shute Barton, Devon, was in a poor state, with rusting *ferramenta* and significant stone decay.

Tile and mortar repair brought the lines back to life; the rusting iron had stainless steel lugs put on and the whole is set up for many more years.

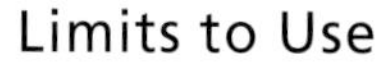

Limits to Use

There are certain applications where mortar repair would not be viable in the long term:

- Water dispersion copings; hood mouldings; channel cills
- Structural replacement
- Mullions; voussoirs; kneelers; paviors
- Extreme exposure
- Spires; finials

This does not preclude its use in these areas but strong consideration should be given for stone replacement in such instances and similar ones. It is best to put aside any thoughts of using cementitious mortars as they are certain to fail by coming adrift, creating capillary actions at the interface, as well as causing the stone to decay and look horrible.

In some cases a lead flashing dressed over the mortar will afford protection to a repair in an unusual/exposed position.

This was presented as finished work by a supposedly experienced stonemason; it needs no explanation to damn it.

Tile Repair

A viable method to repair larger areas of missing stone, moulding and sometimes complete blocks is to build up with larger pieces of terracotta tile or thin slips of stone.

Roughly dress the tiles/slips to the shape of the moulding or with a straight edge and wet them down. Lay them on beds of strong coarse mortar from the bottom up, remembering to slurry up the back of the space and the tiles, tamping them into position to squeeze the mortar up to the face.

This repair can be finished by being left with the edges of the tiles showing; this irregularity can work well with gothic work, perhaps covered with a limewash. Sharply cut edges to stone slips will present better for a neater finish or where a better blend-in is desired. It can also be covered completely by mortar, the joints giving a good key, and the surface worked back to represent stone.

Sherborne Abbey has some lovely honest tile repairs, which stand out yet work with the decayed medieval stonework.

If the repair is long and unsupported, such as a hood mould or mullion, insert a restraint into the stone between the tiles, as common sense dictates; this can be a rawlplug with screw-in wall tie projecting or by drilling two holes angled in to the centre and a hoop of threaded bar slid in under tension lying flat in the joint.

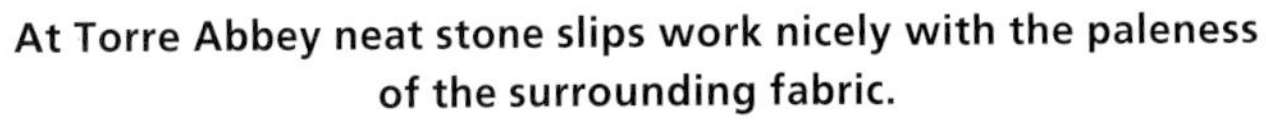

Set in place, the tiles have a mortar build-up to replicate the stone as an alternative to leaving the inserts showing.

At Torre Abbey neat stone slips work nicely with the paleness of the surrounding fabric.

A mortar infill provides a raised surface to prevent water pooling in the dip at the neck of the statue.

A mix of casting and mortar repair. A mould of an existing volute was taken and a lime concrete casting made; this was cured in perfect conditions to make it strong enough as a cast. When fixed, the missing areas were blended in with mortar and then a shelter coat was applied.

The replacement eagle for the Egyptian House (see Chapter 11 for the finished item) was fixed into position and the talons were modelled *in situ* using Roman cement. Having a part with ambiguous dimensions to blend in is a good practice as when a cast is made it may not be positioned exactly in the same place or, as here, the replacement is a different design.

CASE STUDY: COADESTONE RESTORATION REPAIRS

As Coadestone is a hard non-porous material it is not necessary to follow the soft mortar repair guidelines as set out for stone. A previous example showed the fragments of this garland being fixed together and refixed to the building; previous repairs to the missing elements were in ugly coarse cementitious mortar, ineptly made and decaying through.

The missing sections are noticeable after fixing the cleaned repaired pieces.

Basic armature to support and attach the repair.

Sponging back the surface. While this is a bad method for normal mortar repairs, the sheen it gives is a good match for the Coadestone.

Building up in retarded Roman (natural) cement.

Finished repair waiting to dry out and lighten up.

Cutting back to get shape. The speed of set allowed about twenty minutes of working time.

The finished side.

CHAPTER NINE

CONSOLIDATION OF STONE

One of the most contentious areas in the treatment of stone concerns the disintegration and loss of surface material (a process which can be considered 'historic weathering' when the result is pleasing to the eye, or as 'decay' when there is loss of surface material that is unacceptable) and how to stop it. There are many methods of intervention to historic buildings that can be termed 'consolidation', ranging from the insertion of tie or restraint bars within buildings and the filling of voids in masonry with grout through to the rough racking of ruin walls with protective layers.

This chapter concerns itself with the methods of binding together loose particles in the stone itself, as there are many materials and methods, the majority of which have some controversy about their use and efficacy. The main problem with using consolidants of a material that differs from the original is that of incompatibility.

Students carrying out surface consolidation trials to sandstone that is delaminating.

Is Consolidation Necessary?

Promotion of stone consolidation by adding a material into the matrix can be considered a wonderful mix of cod science, speculative reasoning and denial, although often riding on the back of some excellent practical work and research. Pointless claims and dubious wording may be used to sell some of the products involved.

OPPOSITE PAGE:
Romanesque doorway in Glastonbury Abbey, a glorious ruin that is said to house the grave of King Arthur and Queen Guinevere. Visited by over 100,000 people a year it is the focus of many a pilgrimage and faith. Stonework here was consolidated using an organic consolidant – Brethane. (Photo: Kim Coels)

Interpretation of the instructions for proprietary stone treatments would offer that there are only two types of stone in this photograph – sandstone or limestone. With the huge variety of stones used in buildings this is too simplistic, and before anything is applied a lot of questions need to be answered.

But why should we bother? It is the most natural thing in the world that stone will break down over time, otherwise we would be inhabiting a granite sphere rather than our interesting and challenging world full of countless stones; we might also be a bit hungry as there would be no soil to grow anything in. Decay is inevitable even with the most durable stone, yet it is still human nature to attempt to steal a march on the processes of ageing, our precious building materials being no exception, and this has naturally led to the idea that consolidation could work.

What Do Consolidants Do?

In a nutshell, the surface of the stone starts to soften, becoming powdery then crumbling away; to counter this, new material is introduced into the stone to stick the grains back together. This is a great idea and has potential that in very certain circumstances provides a viable proposition for the preservation of historic material where loss of material is unacceptable. Unfortunately it is not a sensible option for general masonry preservation on any scale larger than single stones as it is not likely to perform as well on site as in the laboratory. Once used, it is virtually impossible to re-treat consolidated stone, so the next stage is stone replacement.

Slowly eroding stone becomes friable once cleaned; here the surface is coated with a shelter coat, which is often probably the most effective form of surface consolidation.

Surface treatments to protect or repair stone can bring about problems. Here the wall was coated with water repellant, very similar to a consolidant, which has caused a build-up of moisture under the surface which has then blown off. Note the depth of penetration of the material, shown by the pale edging.

Cautionary Note

The key problem is that if a proprietary treatment (consolidant) does what it says on the tin it is inevitable that the circumstances in which that result was gained are not going to be present on site (remember our maxim: 'stone is a natural material and may differ'), where conditions change as one walks the scaffold. Caution should be the fundamental approach; it could be a reckless waste of time and money to apply the majority of consolidants to a building or monument, not to mention the possibility of exacerbating decay of the stone.

What Could Possibly Go Wrong?

How the surface is deteriorating results in a loss of surface material unique to that particular building – or to be precise, that particular stone – though some generalization of façade decay factors is legitimate. Once the agents and factors of decay have been identified they must be neutralized before any repair is effected – bear in mind that consolidation is not curing the problem, just treating the symptoms. Stage two is where one must decide on what is needed and why (here is where it gets tricky and the concept starts to lose credibility), so to be effective the consolidant must:

- Be compatible with the petrology. The introduction of foreign material into the stone is effectively altering its composition (like for like?), so is the new consolidating material the same as the original binder? As certain consolidants produce soluble salts, more problems could be produced.
- Have an ability to stick the particles together effectively. Some consolidant materials will not adhere to the minerals in certain stones without pre-consolidation chemical treatment.
- Penetrate deeply enough to reach sound material; otherwise the result will be an altered material sitting on the face of the building, the result being just a cosmetic exercise. Consider that even the most insidious materials achieve, at the best, 50mm penetration, whilst most are much shallower than this. Depth of consolidation is usually achieved by dissolving the consolidant in a solvent to carry it into the stone; hopefully the chemistry has been worked out correctly for the setting to occur at the moment of deepest penetration. When the solvent leaves the stone it could pull the consolidant back out with it and as this cannot be checked unless the stone is cut open, it can defeat the object.
- Does it affect the porosity of the stone? It is common knowledge that the prime reason why stone deteriorates is that salts, etc., get stuck in the pores at the surface. It may be possible to pretend ignorance of problems from reduced porosity, but consolidation certainly should not be used if it blocks or restricts the movement of moisture.
- What does it do to the durability of the stone? The introduction of consolidating material in all probability results in material harder than the softer stone below. Will the resulting material have the same coefficient of expansion? If it does not, then the (consolidated) face of the stone may slough off.
- In some circumstances the consolidant can alter the appearance of the stone at the time of application or it can be affected adversely through weathering/ pollution and turn a different colour from the stone over time. Obviously this is an undesirable effect that is impossible to rectify once the project has finished.

Sandstone slab, one of many delaminating in this burial ground; the repair specifications gave a method for consolidating these based on limestone. If there are any doubts about the method used or it is felt that a mistake has been made, you may get a cool reception when raising issues, but nowhere as bad as if the stone is destroyed by blithely carrying out incorrect treatment.

All in all, there are a lot of things to be considered before consolidating any part of a building; as in all applications of extreme treatments the results, if not prepared for, can be detrimental to the stones' appearance at least, and destructive to the rest of the surrounding material at worst.

Not the Last Word

The Venice Charter is one of the defining statements on conservation ethics and interestingly gives hope to all those peddling consolidants as the elixir of life for stone:

> ARTICLE 10: Where traditional techniques prove inadequate, the consolidation of a monument can be achieved by the use of *any* modern technique for conservation and *construction*, the efficacy of which has been shown by scientific data and proved by experience.
>
> (The Venice Charter: International Charter for the Conservation and Restoration of Monuments and Sites; *II International Congress of Architects and Historic Monument Technicians, Venice, 1964*. Adopted by ICOMOS in 1965.)

Sadly, unscrupulous practice could advocate the first part whilst ignoring the last part. When used appropriately,

THE LIME METHOD

There is a small ray of hope that shines for owners of powdering limestone buildings – the lime method. This is a minimum intervention technique that relies on the anticipation that lime dissolved in water can be easily applied to the crumbling surface where, exposed to the atmosphere, it transforms back to calcium carbonate (the natural binder of limestone) and glues the grains back together. This has caused more debate in this country than all the other methods combined – with the jury still out. A common argument for championing this method is that if it does not do any good at least it does not do any harm. Considering the following figures may be enlightening:

- Calcium hydroxide in solution (limewater) contains approximately 1g of lime per litre of water.
- Visualize 10 square metres of ashlar crumbling to a depth of 10mm. This will give us 10 litres of crumbling stone with a weight in excess of 20kg.
- Assume the loss of binder is 25 per cent and this needs to be replaced to consolidate the particles.
- To facilitate this we need to reintroduce 5kg of lime – logically achieved by applying 5,000 litres (that is five tonnes) of limewater.

This presumes everything works famously and that problems do not occur, such as carbonation blocking the pores, water evaporating off, or some busybody pointing out that as most problems in stone can be traced to the presence of water, the addition of five tonnes of the stuff into the fabric of a historic structure may be detrimental.

More salient, though, is the fact that the drive behind this method came as a result of research, by placing decaying stone in a bath of limewater. Regrettably it was found that no calcium carbonate was actually deposited in the stone. Even using crushed limestone for the trial did not yield any better results; yet this was, and still is in some quarters, mooted as the best method of 'minimum intervention' consolidation for friable limestone.

The use of nano-lime is now being explored for fine penetration of limestone. At the moment, from all the research (the limit of my knowledge so far) there does not seem to be any consensus on whether it works, has any benefit or is practicable. Until I have had a go with it, I will not be writing about it.

The college in Venice where I was first introduced to the joys of consolidation. The plight of the sinking city was one of the driving forces behind the ratification of modern conservation.

consolidation of stone can be an effective tool in the preservation of our built heritage, but it must be ensured that the last repair technique that happens to a stone is the result of serious thought and good practice.

Consolidants should be used wholesale on large stone structures only after an appraisal has been made which considers the risk involved, the benefits to be realized, and the probability of loss of the material without intervention.

Replaced stone had been cleaned of dirt. The alkaline grout had mobilized the iron pigments and brought them to the surface; this is not a long-term problem as the stone will bleach back in time, but in the meantime it looks unsightly.

This rich Romanesque arch in Glastonbury was consolidated with a surface-applied mix almost thirty years ago. While it has preserved the detail the colour has changed; compromise is often a necessity if the pros are to outweigh the cons. (Photo: Kim von Coels)

Primary and Secondary Requirements

In addition to creating new bonds, a good consolidant should meet performance requirements concerning durability, depth of penetration, effect on stone porosity, effect on moisture transfer, compatibility with stone, and effect on appearance. These primary requirements are considered to be generally invariable; that is, they are essentially applicable to all stone consolidants regardless of the specific use.

Secondary requirements may be imposed in addition to the primary ones because of specific problems encountered in certain structures. For example, to require a consolidant to immobilize soluble salts in a stone would be a secondary performance requirement.

Types

Chemical and physical properties are the identifying properties of the four main groups of consolidants: inorganic materials, alkoxysilanes, synthetic organic polymers and waxes. Selection of what material to use depends on many factors including the type of stone to be consolidated, the processes responsible for the deterioration of stone, the degree of stone deterioration, the environment, the amount of stone to be consolidated, and the importance of the stone structures. As a universal consolidant does not exist, the preservation of each stone should be tackled as a unique problem – which tends to make this an expensive and time-consuming exercise.

Tyndall stone cornice was exhibiting hairline fracturing, so tie-bars of our own design were inserted; here the core drill is used with a two-metre bit to make the hole.

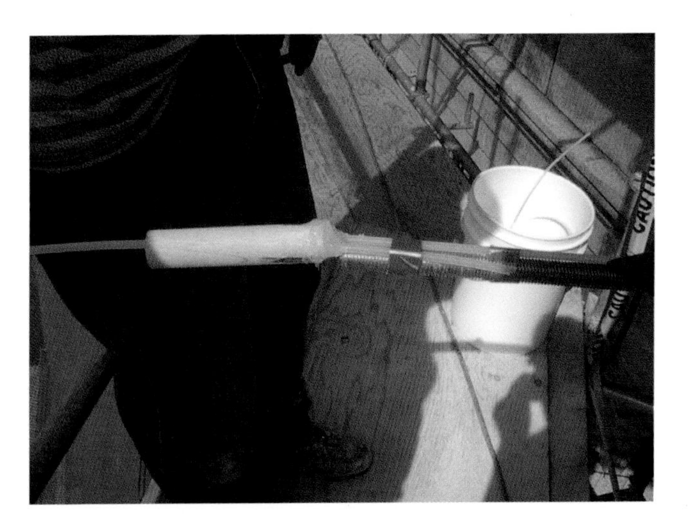

The bar had a feed tube to its farthest end and an outlet to let the grout feed back out once the hole was filled. The wooden plug was sealed in to prevent spillage then drilled out and a core plug of stone fixed in the hole.

Pressure-filling the anchor holes with a resin grout; the outlet pipes are above the cornice so they can be monitored for fill levels. The sprayer is to keep the surface of the stone damp so that if any resin leaks out it will not stick to the stone.

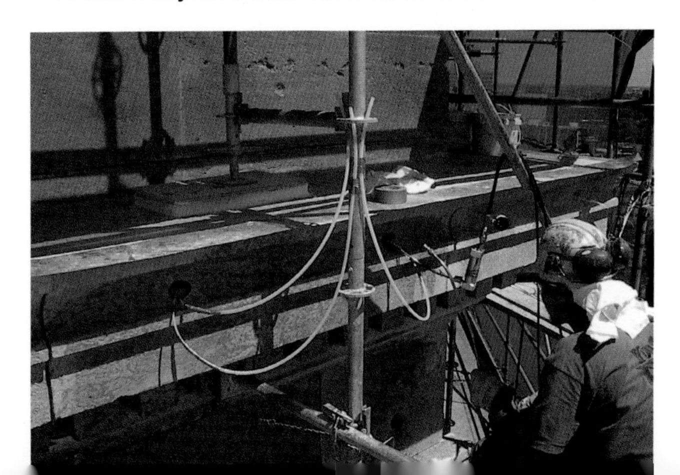

CONSOLIDANT CHECKLIST

This is set out here as a table: the aim is to answer as many as possible.

<table>
<tr><td colspan="6">Material's properties</td></tr>
<tr><td colspan="3">Consolidants
Correct viscosity
Correct surface tension</td><td colspan="3">Stone
Porosity
Dust-free
Humidity</td></tr>
<tr><td colspan="6">Consolidation process</td></tr>
<tr><td colspan="3">Setting time:
Slow
Rapid</td><td colspan="3">Changes during setting:
Does it shrink?
Does it react with the stone?
Is heat involved?
Does it change the nature of stone?</td></tr>
<tr><td colspan="6">Consolidated stone</td></tr>
<tr><td colspan="2">Strength:
Too strong
Too weak</td><td colspan="2">Reversibility:</td><td colspan="2">Appearance of the stone:</td></tr>
<tr><td colspan="6">Will it last?</td></tr>
<tr><td colspan="6">Long-term changes:
Will it shrink?
Will products of decay affect the consolidated stone?</td></tr>
<tr><td colspan="2">Cohesion:
Will it lose cohesion?</td><td colspan="2">Reversibility:
Will it become insoluble?</td><td colspan="2">Appearance:
Will it change colour?</td></tr>
</table>

It would be easy to embark on a major discussion on consolidants but this would be outside the scope of this book so we will leave aside the intricacies of consolidant lore and the science of consolidants to focus on how they are used and applied.

Testing

The first task is to check that the stone will absorb enough of the consolidant to make a difference; therefore some testing is necessary. It is obvious that the stone must be porous, so how can we check this?

A porosimeter or Rilem tube should be fixed on vertical surfaces using plasticine and filled with water or the liquid consolidant. Cleanliness is essential, as dust can block the absorption and prevent the porosimeter from staying fixed, so the surface should be as dust-free as possible; blow it clean with pressurized air, but not so violently that it disrupts the stone. The stone must be protected from the oils in the plasticine, so attach a piece of tape cut to the size of the porosimeter opening on the area then paint a water-soluble PVA (so it can be washed off later) around this beyond the edges of the apparatus, then remove the tape. Knead a lump of the plasticine until malleable and form a ring on the PVA, pressing the porosimeter into this firmly until it holds. A bit of gaffer tape around this onto the stone will give a bit of back-up – this could also secure some absorbent paper at the bottom to soak up if a leak happens; leave everything for a while to firm up, then pour the liquid into the tube.

The disappearance of the liquid will indicate it is soaking into the stone. (Volatile liquids should be covered to prevent evaporation loss but do not make a vacuum.) Keep a check on the time it takes; now rough maths (area of stone × volume of liquid) will give an indication of how much consolidant solution you can expect to put into the stone.

Rilem tube stuck onto the stone to test porosity and liquid take-up. Liquid is flowing into the stone and wetting up the surrounding area; the stone had been consolidated in the past and it was impossible to introduce any pertinent amount of consolidant.

Did It Get In?

The solution of consolidant in a medium has to flow into the stone, and then the medium comes back out (boils off), depositing the binder inside the pores. The main argument with this is: if it can carry it in, surely it can carry it out again? To confirm the efficacy some accurate measuring must take place: weigh the plasticine and porosimeter, place an exact volume into the porosimeter, and have an exact consolidant/ medium ratio.

Once satisfied that the stone has absorbed as much as it can, let the medium evaporate out of the porosimeter. The consolidant that has not gone in should be sitting in the bottom of the porosimeter. Now place a plastic tub (of known weight) flush underneath, slide a flat-bladed scraper in under the plasticine and flip the whole lot into the tub, making sure all the liquid is collected. Weigh this, subtract the equipment weight to find the weight of the remaining consolidant and subtract this from the original amount used. You now have a figure to work to and this will form the basis on how much to actually use.

The previous consolidation treatment had worked fine on the surface, but the decay of the stone continued apace behind this crisp layer, with the result that all the detail is being blown off and a powdery surface left. Some of this has been consolidated with a shelter coat, which will at least wear out quicker than the stone.

Application

Before any work is undertaken, read the material safety documents to know what is being used and its effects on you as user and on the environment. Always wear the appropriate PPE and for noxious fumes have full-face masks with the correct filters in – if you can smell the consolidant while wearing one, it's got the wrong filter in! Disposable gloves come in a variety of materials so choose the one for the job. Also be aware that some people have an allergy to latex so always carry a variety of types; good quality is essential as the cheaper ones will tear easily; the difference in price will be negated when you need to use half as many.

BENEFIT OF SWEATY PALMS

Ingrained dirt and resin can be difficult to remove from hands without serious scrubbing – never use solvents to do this! Heavy-duty hand scrubs with abrasive (but gentle) particles, of the type found in engineering and mechanical workshops, are the best for tough staining. The usual hand-wash found everywhere does not tend to have the scrubbing power needed, though it can be improved by adding a spoonful of sugar to provide some abrasive action – this is harmless and easy to do. Experience has found that stains will 'sweat' out of the skin when disposable gloves are worn for any length of time, making it easier to get pristine hands – this is not for everyday hygiene, but it does work.

Mess

Spillage or runoff should be prevented by adequate masking of the surrounding area; plastic sheets below can be manoeuvred to funnel into a container or taped into a gutter. Make sure the sheet materials are impervious to the chemicals in the consolidant, as a spill from a build-up of sticky liquids is hard to contain and will probably catch everybody unawares, exacerbating the problem.

Putting It In

The consolidant needs to be in solution, and should remain so at all times. If it is a slow-dissolving mix, prepare it well in advance and ensure full mixing before use. It may need to be

Catch trap and guttering for spills from grouting on the lift above. The design allows any leaks to be washed off the stone and channelled away safely.

in motion regularly to prevent separation – know the material. For small amounts a heated magnetic stirrer will get it into solution; make sure that the beaker is capped to prevent evaporation of the solvent.

A heraldic shield from the Egyptian House. Removed in many pieces it was rebuilt and cased in plaster, the fine joints vacuum-consolidated with resin and the back reinforced with mesh.

A mix of natural cement and LECA was put in.

The back was filled with the concrete mix which was levelled off.

The shield out of the case, painted, and being fixed in position back on the building.

Facing up is used to hold delicate or fragile objects together while moving or working on them. This is simply accomplished by layering sheets of strong paper towel (shown here) or cotton fabric soaked in a solution of water-soluble PVA and water, then tamping it in and then letting it dry; you can use a hairdryer to speed this up.

Brushing On

Good-quality brushes are best here; those used for varnish will hold and place materials evenly, so try a couple of types to see what works best. When finished with them, clean and store them in a big jar with some thinners/white spirit or acetone (though this can destroy the glue holding the bristles in) in the bottom.

Decant enough mix to use in a short period of time and seal the rest until needed – always mix before pouring. Load the brush and start from the bottom, working upwards in vertical strips at a speed that lets the consolidant sit on the stone but not run down. As the objective is to get as much in as possible, recoat as the consolidant is absorbed; do not let it dry, as this will prevent further applications moving in. The medium/solvent will start evaporating off (as this will be relative to conditions, a windbreak may be needed) so there is an issue here: the consolidant must be deposited and start making bonds, so the solvent should move out, but if it is still in solution then it can be drawn out again. If necessary, slow down the evaporation by taping a plastic sheet over the stone – this can be attached to the masking tape around the edge to make a seal. Pinpricks in the sheet will allow gradual evaporation. The manufacturer's information should be of use here, with the proviso that their tests will have been done in perfect conditions, so adapt and experiment before using on the precious stone.

Spraying

For very small amounts it may be possible to use a hand sprayer, but solvents can attack cheap seals, causing mess and loss of pressure, and it will be difficult to apply evenly; before going on site try the sprayer to see if it can cope. Spray the dilute solution on using an airbrush for features; a standard paint sprayer powered by air for bigger sections will give even coats and controlled build-up. Airless sprayers can be found at garden centres or DIY shops and are the preferred method for those not using compressed air (but try to get one with metal components). Practise to get the settings right, and spray onto a spare bit of stone or clean wood to check the output and spray pattern. Have spare nozzles in the kit and clean all

THE MUSBURY REREDOS

Attached to a damp wall in the church, the beautiful mosaic reredos panel had developed a large bulge in the centre. As it looked extremely thin and we did not know what was at the back, it was faced up with PVA and paper, with a plaster case put over this to reinforce it when it was lifted off the wall. At the workshop it was found that it had been built on a zinc sheet in a ferrous frame (attached by protruding brass rivets), which was rusting badly, threatening more disruption and collapse. This frame was carefully cut off using a Dremel diamond disc, leaving the rivets in place, and the zinc cut away at the back of the bulge. A stainless steel frame was attached and mesh was wired onto the rivets. This was later filled with a GRP layer to tie everything together, but first the bulge had to be sorted out. The tesserae were set in a linseed oil mastic, so this was softened in the area of the bulge by a poultice of dichloromethane paint stripper and warmed by a hot air gun. The bulge was then pressed back to line up properly. It was cleaned and faced up again before being fixed back in the church; the facing up was washed off with warm water and then given a microcrystalline wax. The result, to put it mildly, was stunning.

The new frame in place; the mesh is wired onto the rivets in sections, allowing access to the cutout behind the bulge.

The rusting frame at the back of the panel, where the rivets are starting to stand proud of the corroding metal.

Facing up prior to transporting back to the church.

The finished project, with a new piece of masoned alabaster to the surround.

the parts after work with a waste spray using solvent only. Be aware this will put a lot of (generally toxic) vapour into the air at the workplace so, once again, remember PPE!

Poultices

Applying a consolidant in a poultice is probably the least effective method available. Mixing a paste that contains the consolidant and applying it to the wall is fraught with questions, the obvious one being why the liquid would leave a nice stable environment like the poultice to enter a much less absorbent place. Also, how is this checked?

One rule when making any paste is that you always add the dry material to the liquid, never the other way round. Combine the dry poultice materials together, mix together any liquids and then start pouring the latter into the tub until a suitable paste is achieved, using either a plaster mixing bit in a drill or a non-metallic agitator (such as a wooden paddle).

The stone and its surface must be wetted up to dispel surface tension and the poultice applied with a trowel, pressing it hard into the stone. For large areas a plasterer's large trowel can be used; start at the bottom and push on with a sweeping circular motion. This is only good for flat surfaces and has the problem that working the poultice will draw the liquid to the outside – so do not overdo it. Cover with a rubber or plastic sheet (that will not be dissolved by the solvent) and let it soak in. Dispose of the used poultice according to the manufacturer's specifications.

Outer bailey wall of Corfe Castle prior to coming under the care of the National Trust. The level of the ground had increased and spread over the fallen blocks of masonry; there was also 1950s infill of the gaps in the wall.

After excavation, I capped this exposed core of the fallen castle wall with rough stone and flint using lime mortar in a style to suggest the layering of the rubble infill, so viewers could identify the orientation of the block. To remove the modern infill and capping on the wall, it was specified to use hammer and chisel; this was so painstaking and impractical (we met one of the men who had built this and he told us of the quality of the hydraulic limes they used, which we seconded!) that hammer drills with chisel bits were the result.

Vacuum Impregnation

Essentially a museum process, this is difficult to carry out *in situ*, the obstacles being selecting the area and isolating it from the rest of the structure. The objective is to draw all the air from out of the area to be consolidated using a vacuum pump and then letting the consolidant be pulled into the voids; it is possible to do if you can hermetically seal the area being consolidated.

Setting up vacuum impregnation tests in the workshop; the film is being sealed around the edges.

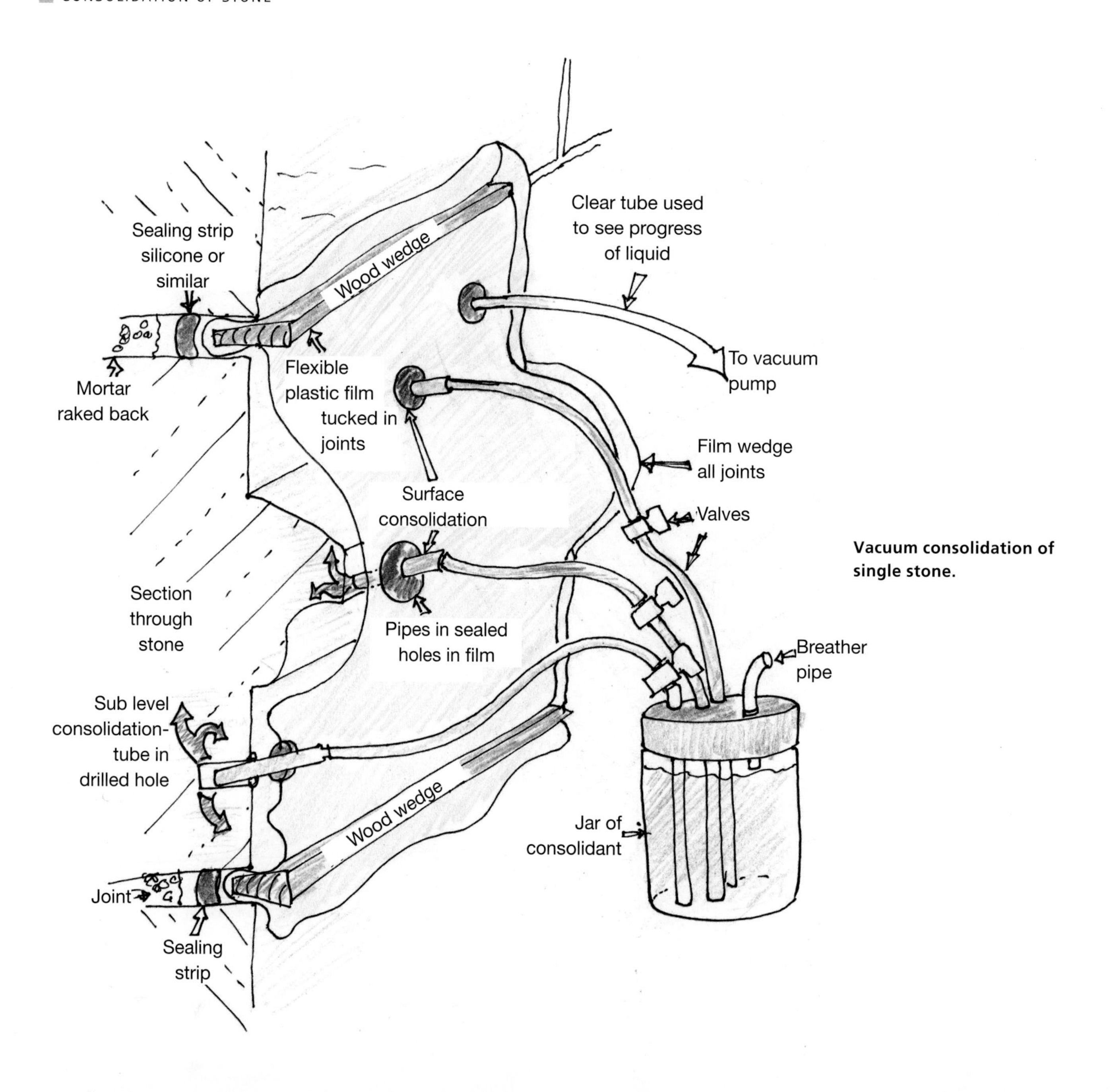

Vacuum consolidation of single stone.

If it is possible, go to the joints in the stone and clean them out thoroughly, then paint them to depth with a water-soluble coating of PVA. The surface should be as cleaned as possible of dust and loose material, but not to the extent of losing material that needs to be saved. Cut the membrane sheet to allow a good tuck into the joint, loosely align it and mark where the nozzles for pumping and consolidant are to be. Put the nozzle apertures into the sheet and seal them inside and out. Now apply sealant by mastic gun into all the joints, place the sheet and poke the edge into the joints; if possible tap in a strip of wood or similar over the hem to wedge it in place. Let the sealant cure before applying the vacuum.

Do not attempt to do larger areas than can be cleaned effectively in one hit. Mount the consolidant reservoirs securely, either clipped to scaffold or screwed to wedges of wood in joints, join them to the nozzles (do not forget to install taps/valves in the pipes) and fill them with liquid. Connect up the vacuum pump (having remembered to put a valve in the pipe) and draw the air out until a high vacuum is reached, then close the valve and open the consolidant taps. The liquid should be drawn into the voided pores of the stone without collecting on the surface, so control the flow. Once satisfied with the impregnation, give enough time for the consolidant to start gelling then quickly remove the sheet, mopping up any drips. With the solvent, remove any surface accretions of the consolidant, lightly spraying the solvent and swabbing off, or by pressing absorbent cloth onto the surface and brushing with solvent to soak up the residue.

Some heavily textured or open-pored stones may be hard to clean off by this method, so to reduce the amount of consolidant on the surface, drill shallow unobtrusive holes of the same diameter as the tubing, at angles into the stone. Clean these out with a water/alcohol mix and blow with compressed air. Use a syringe with solvent or alcohol to test the effect by squirting into the hole (make a seal with tubing or plasticine); the area around should darken up indicating a spread of liquid. The tubing from the reservoirs should protrude past the membrane, and be positioned to fit into these holes so that the consolidant is delivered behind the surface. Apply a sealant ring around the tube and press them into place, then carry on as above.

Medieval statuary is probably one of the best uses for consolidation – but for it to really work they should be treated in the workshop to allow greater control of the processes. These statues are often designed to be removable, as due to their significance they had to be kept in tiptop condition, so it is worth looking to see how they are fixed. A tale is told of 'conservators' who wanted to drill through the front of some important statues, then across the gap and into the building behind to insert a tie-rod. Luckily a stonemason colleague pointed out the sheer damage this would do, and, more significantly, that the statues were attached by a hook through a loop of metal hidden in the niche, and could be lifted out easily.

STRANGE BUT TRUE!

In the infancy of organized stone conservation, many strange things happened, but one that takes the biscuit was when one of the great cathedrals of England was being worked on. Precious medieval statues were wrapped in a poultice of quicklime, and then saturated; this was a gratuitous and dangerous low point in our work, considering the resulting destruction of irreplaceable artefacts by such an inept and farcical method.

The 'pioneering' expert involved with this, while eschewing the methods of the stonemasons in the yard as inappropriate to his great work, was often observed by the said stonemasons surreptitiously 'borrowing' tins of polyester resin adhesive to secretly repair stonework and statuary broken in the process of its 'conservation'.

Hard-capping of core is one of the two options used for the protection of historic masonry ruins where the core is exposed. At Corfe Castle the walls were hard-capped and then soft-capped in the same place. Here the large mass of mortar is being applied, quite wet and trowelled smooth; the next stage is that a system is fixed to provide a growing medium for the grass.

Soft-capping/consolidation is a simple method that is becoming more popular. It involves creating a living cover of turf over the top of exposed masonry core in ruins; this creates a pastoral and romantic effect to these structures. The idea is that the grass absorbs water and binds the loose material as well. Obviously though, other plants and animals will start colonizing the sod and problems can occur, which is the reason that all the original growths that spread over ruins were so assiduously removed many decades ago, and now we come full circle.

CASE STUDY: GARGOYLE RESTRAINT

This rare Green Man grotesque was a gargoyle (French: *gargouille*, meaning gullet, as they act as waterspouts though their mouths) on Shute Barton and was in danger of parting company with the building; the large open vertical joint to the quoin stones was an indicator of movement outwards. It was impractical to take it down and rebuild so a restraint was designed and put into place.

Stainless studding (threaded bar) fixed through the quoin stones, jumping the joint; these were set in resin with the added restraint of nuts and washers set in counterbores, which were left uncovered as this was high up.

Restraint fixing in the throat of the gargoyle, the studding going up through the head; at the sides are angled holes that supplementary brackets will fix into and be attached to the main bar.

At Shute Barton, the corner of the building shows the widening joint.

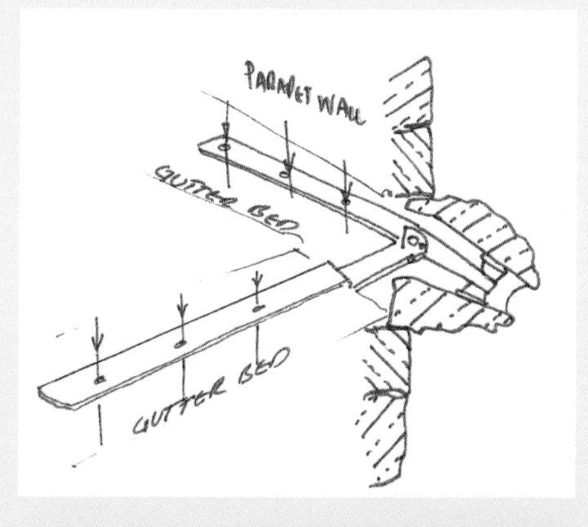

Sketch for the restraint design.

Studding projecting above the head, with tie-back bar slid on for sizing.

Forming a protective lead cowl for the stone. The bar projecting to the left is the end of a Y-shaped strap that is anchored into the floor of the parapet gutter; these were sealed into the gutter with hot poured bitumen.

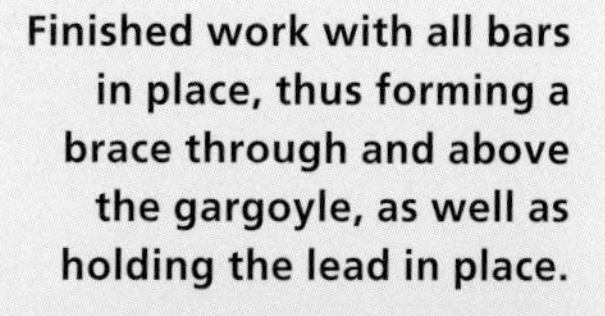

Finished work with all bars in place, thus forming a brace through and above the gargoyle, as well as holding the lead in place.

CHAPTER TEN

CLEANING STONE

This subject would be worthy of a complete library in itself, were we to go into the reasoning and science of removing dirt (soiling, biocolonization, sulphation – anything that, in the experts' view, should not be there). As a practical book it is not necessary to cover all the academic analysis, but before the techniques are covered some theory is essential. Here is not the place to discuss the dilemma of cleaning buildings (and neither will these issues be discovered on the label of proprietary cleaning materials), but it certainly is the place to arm you, the artisan, with an approach (and having the right approach is akin to having the right chisel) for finding out exactly what the situation is before making a plan.

Controversy

The cleaning of masonry façades is significantly the most visible aspect of 'conservation' work carried out in the built heritage sector; it is certainly one with capacity to cause immediate and future damage when carried out incorrectly. Sadly, all the evidence of inappropriate removal of dirt and stains still does not deter unqualified and inexperienced attempts to continue damaging our built heritage.

OPPOSITE PAGE:
The staining of stone can be significantly affected by simple changes in the situation; here a fountain in Aix-en-Provence shows algal growth caused by moisture and soluble salts, while metallic salts keep growth off but add their own stain to the stone.

Why Do It?

Before any cleaning takes place, questions should be answered to the satisfaction of all. The major one is: why clean? Is the cleaning going to benefit the building and, more importantly, is it necessary? Good maintenance will keep a building free from the accumulation of dirt and allow natural weathering to proceed, whereas abrasive or corrosive actions of many cleaning methods can remove the protective surface ('patina' for

The conservation of the Coliseum in Nîmes, the best-preserved Roman arena in France, is an ongoing project including gentle cleaning of the façade. The environment is not so affected by industrial pollution here, so the staining is not as great as in other cities such as Paris or London.

This stone window jamb obviously needs cleaning, but at what point is enough? Is the graffiti unwanted damage or should it stay as historical evidence? If the stone is lightly cleaned it will actually highlight the scraped-in names, altering the appearance. Conservation decisions are never simple!

the estate agents) formed on masonry, with the unsurprising result that a clean building does not stay that way!

Note that decay and soiling can gain a foothold on a freshly cleaned surface far more easily than on one with its own collection of dirt.

Incidentally, never consider putting a protective surface coating on masonry; it is probably the worst thing that can be done to a building. Do not apply anything – it will survive a bit longer.

Unwashed and Unwanted

The decision to clean demands knowledge of the dirt and of the material it is attached to, and of whether the removal of this material will affect the masonry; so there must be identification of both the unwanted material and what it is attached to. Even the most monolithic building will have a mortar different from the stone, whilst many historic buildings can have a variety of materials used over the years. Due to the huge variety of stones, bricks and other materials used in buildings, and the wide range of environments in which they stand, it is obvious that one type of cleaning will not suit all.

A veteran stonemasonry teacher wrongly identified the stone of this monument in Bournemouth and proposed aggressive cleaning. Luckily we managed to inspect and comment. The subsequent cleaning with gently sprayed water and brushing has prevented damage and the results are lovely.

Recipes

There will be no formulas given here because the needs of any intervention to historic material, especially stone, dictate that the procedures set out below will need to be undertaken to determine what is going to be used and how to use it. Many books on this subject have been produced that contain specific recipes to deal with the evils of masonry soiling; as a guideline and to learn what has been used for certain situations, they are certainly worthwhile. Unfortunately, people without understanding of this work take these as gospel, learn

Probably the most extreme work I have been party to was carried out in the name of conservation, here in Cirencester. The stone surface of the façade was deteriorating badly and the proposal to take it down was going to be impractical and costly. The drastic alternative had the entire surface ground back to sound using a massive milling machine and hand tools in the tight bits; much new stone and mortar repair completed the task. There was the usual hullabaloo: the repair work was extremely good, and, as the stone was all freshly worked, it could not be easily discerned. Interested groups were shown around by an architect from one of the statutory organizations in heritage, who, as he could not spot our work, decided (erroneously and without discussion) that the new stones had all been put in edge-bedded, indicating to his audience the lack of skills in modern workers. It was not a jolly discussion that followed.

them by rote and insist on using them for all manifestations of similar problems, occasionally getting lucky, but generally blinding the uninformed with their (limited) science. Do not go down this road!

When figuring out what to do, start from the beginning and use the available science and experience to formulate appropriate methods; remember that all stones can have unique characteristics, even those from the same bed in the quarry. Add to this varying environments, localized agents of decay and a mixture of positions in the structure, and 'one size fits all' starts to look like a dangerous policy.

The Prize

Clean masonry is perceived differently by everyone: owners and investors want pristine stonework to show value, surveyors want the ease of 'reading' the construction, conservators want removal of harmful material, and contractors and developers need to show investment and work done.

Some of the participants may need to be reminded that there will be some situations where the most adept cleaning methods are simply holding measures and that the original design of the building will continue to re-create the problem.

Value and elegance is added to buildings when they are clean. The Royal Crescent in Bath is a wealthy area and most buildings have been kept clean, so the build-up of scurf on the stonework is markedly noticeable when there is a pristine house in an otherwise uncleaned row.

Run-off from the portico roof is promoting the growth of soiling algae, and the beginning of lichen colonies, over specific parts of the stone creating unsightly soiling. While this (the symptom) can be cleaned, the real task is to prevent the cause by addressing the roof design.

Before You Start

Weathering of stone can be seen as providing character to a building, and accumulation of layers of soot or surface treatments can ruin the aesthetic quality of the structure; so there are several good reasons for cleaning. The term 'desired result' is crucial, meaning the level of cleanliness that has been approved by those responsible for the project outcome. The

Routine maintenance cleaning of the marble statues on the Albert Memorial using pressurized steam; if only all buildings could be cared for on a regular basis!

Here in Nîmes, cleaning of an old school building is carried out with a modern version of the traditional sandblaster; with the addition of a water feed it is not too aggressive.

measurement of how much cleaning is necessary and what methods are to be used for this should be established before the project goes to tender, as it should be the contractor's role to fulfil a requirement rather than set a level. Stone should be cleaned using the least aggressive methods to achieve the desired result. Pre-project trials are compulsory for work of this type, and should be undertaken by skilled and experienced conservators; the result will then be a standard by which all the subsequent cleaning on a large scale will be calculated. As we saw in Chapter 2, the conservator carrying out the trials will be obliged to compile a report covering all aspects of the building, material, tests carried out and recommendations. Before any programme of cleaning is undertaken or even specified, all those involved should agree to ensure that the method used is appropriate, non-destructive and practical.

How dirty the building is may seem obvious to the man in the street, but the point here is one of defining what the unwanted material is; it is necessary to know what one is dealing with, what it is doing and how to get rid of it.

Cleaning projects prescribed solely on the appearance of the building and aesthetic considerations of the client instead of by investigation and fact, are not going to be effective; before the scaffold goes up certain criteria need to be satisfied.

Setting up a trial stone-cleaning sample to remove verdigris staining. Two panels are used here, each with a different dwell time to compare the effect of leaving the poultice on longer.

The Masonry

What is the material of construction? This may seem obvious in the masonry world but those with experience will agree that in this rocky little isle there is a huge variety of stone used in construction. Historically it was not unusual for any (handy) materials to be used, be it from a field, quarry, tumbledown building or even ballast from the bottom of a ship. It can be disastrous if the 'right' cleaning agent is applied to the wrong stone.

The Dirt

What is the dirt? Remember the old maxim: 'dirt is material in the wrong place'. This is crucial, as the correct processes of examination to ascertain the dirt's true nature can incorrectly

The wrong stone: the Palace of Westminster was rebuilt in the nineteenth century using Anston magnesium limestone, because it was a good colour, cheap and readily available. However, due to the poor quality of supplied stone, it suffered badly from pollution; by 1950 it had been replaced with Clipsham.

identify many types of soiling. If the soiling is biological, could this be by a protected species (such as many lichens) or a disfiguring colonization that is actively consuming the stone, possibly providing the foundation for more staining? The natural weathered patina of (bare) stones is probably their best-loved feature, caused by gentle polishing of the surface by the elements; cleaning with even the mildest abrasive can remove this, and once the surface is dry it will be cleaner but dull. Is this what is wanted?

Origins

Where does the soiling come from? Discerning whether a material from within the stone is leached out by chemical reaction or is an accumulation on the surface by airborne deposition can dictate the approach. The building may have faults in design, or may suffer a lack of maintenance that can create problematic environments encouraging the accumulation of soiling. The umbra of trees next to walls can become moist greenhouse areas that encourage algal staining. Is the dirt harming the building or is it a symptom of some other process of decay or neglect that needs to be addressed before an effective clean can be made?

A moist environment and the shelter of trees allow biocolonization of Hamstone memorial stonework.

Here a contractor's error has produced a need for cleaning; oak laths had been stood on end on the floor and the tannic acid had leached out, causing the brown staining. To get this out a poultice of limewater was used to help neutralize the area and then hot soapy water lifted the sap.

How Hard?

Whether the dirt can be removed without harming the fabric of the building is probably the most important of all these questions, bringing together all the other strands of

Rather than wash and scrub the old copings, the so-called restoration work in Evershot, Dorset, included wire brushing of the surface, leaving myriad scoured lines. Note the failed mortar repairs, where the contractor slapped a ready-mixed mortar on, leaving the unprepared fills to fend for themselves.

interventive work. Conservation principles dictate that the method used should be the least aggressive to reach the desired level of cleaning, whereas the contractor may wish to use the strongest available method to ensure the cleaning gets rid of everything quickly and efficiently (that is, cheaply).

Masonry provides a porous surface for dirt to inhabit, unless this dirt can be persuaded to come out (by poulticing or washing); to physically remove it will inevitably involve loss of stone to some degree.

Not Too Shiny

How clean should the building be? If the building is cleaned too aggressively it could cause disharmony in a townscape of other dirtier buildings, whilst leaving a grubby façade will cause the same effect. Once the criteria have been satisfied the project team should be introduced to a sample of cleaning that must be agreed on and adhered to for the whole term of the project.

Will the effect of the cleaning last and is it safe? As mentioned, cleaned buildings provide a better foothold for new soiling, so the cleaner it is the faster it will become dirty again without due maintenance.

Keeping It Clean

One of the biggest problems is that the materials used can harm the stone over time, causing minerals to migrate, change colour or disintegrate. The best tool to counter this is communication; look around and ask questions of other practitioners and heritage bodies – conservation is dedicated to education and progress through the dissemination of experience and information, so there should be no obstacle to obtaining records of previous cleaning, the materials used and the effects during and afterwards.

Water is the most widely used cleaning agent, rightly so as it is extremely effective; conversely it is the force behind the majority of decay problems in buildings. It would be a poor result were the building to be clean outside and rotting inside.

Summary

Buildings will get dirty and they may need cleaning if this is detrimental to the structure or it detracts from the aesthetic statement. Before diving in to save the day, investigate, identify and understand both what is happening and what should be done.

Cleaning Techniques

The three main methods for stone cleaning are water-based, chemical and mechanical, with many areas of overlap between them. Here they will be demonstrated in semblance of order pertaining to aggressiveness, but do not get confused if techniques involved seem dissimilar to the title; it is safe to say that everything is connected, just not always in the most obvious way.

For the purposes of this book, the hypothetical materials to be cleaned are areas of vertical stonemasonry that need cleaning; the following exercises really fall into the area of trials, as they would need sensible scaling up for larger areas.

Mild and Gentle

Getting rid of loose dust is the first activity. Brush down with a soft bristle paintbrush, working from the top and tracking

Any chance of mess from work should be minimized by adequate protection; put down boards on top of plastic to protect floors that will be walked on.

across in strips to get the debris going down. Do not wipe the brush as in painting, but get the end of the bristles in contact with the surface and dislodge dust with a rotating or flicking motion. Progress from this in stages with various types of soft to stiffer bristled brushes to remove more tenacious dust, stopping with the one that is effective. For awkward nooks and crannies, make up wooden scraper tools that can dig out accumulations of muck without scratching the stone.

Scraping

Thick clumps of moss and lichen will absorb water before it hits the surface; these need to be lifted off physically to allow water or brushing to get past. Make up wooden spatula scrapers that will not damage the stone or use plastic windshield scrapers; do not be tempted to try a metal blade as this will scratch or polish the stone.

Non-metallic scrapers must be used: wooden sticks (plastic can be used as well) can be cut from any spare wood and shaped for specific soiling and areas.

METAL TOO HEAVY

Occasionally some barbarian will attempt to clean up stone using a standard wire brush, the result being hideous to see, as it leaves a legacy of tiny morsels of metal embedded in the surface ready to rust and stain; this should never happen with a conservation project, or on any stone come to that. Unhappily some feel it is okay to use a wire brush on stone as long as it is non-ferrous; this may mean a stainless steel brush so tough that it will scour the surface leaving trenches that are an absolute boon for dirt and flora to notch into, leading to rapid accumulation. Alternatively, brass wire brushes are mooted; these, if anyone has bothered to look, leave a metallic sheen on the stone (where do people think all the worn off bits of metal go?) that will react with acid rain and put salts into the stone, as well as leading to verdigris.

Never use treatments without knowing what is going to happen. Notice my 'get-out clause', as it is often the case that one person or another will proscribe one of the more interesting techniques, but as long as the correct protocol to the intervention has been taken and indicates this is the best way, then pragmatically it should be so.

Air

Compressed air using a blow gun works well because the pressure can be ramped up to dislodge more tenacious dirt, but be aware that it can also lift flakes of stone and erode mortar, so always start on low pressure. Work in horizontal sweeps across the surface with the nozzle pointing generally in the down direction, bringing it more horizontal to worry out ingrained dust, and then progress down the face. Try this in conjunction with brushing or picking by wooden tool.

Washing

Using water is the most obvious way to get dirt off and, wonderfully, it works incredibly well in a lot of cases. Whenever cleaning with water is done it is better to have a live supply as this is essential for rinsing off effectively; it is pointless to scrub a surface clean with buckets and brushes only to have the soiling running down and sticking to the lower stonework as it dries off.

First stage in cleaning. Water and brushing will often surprise with the quality of the result.

Water floods across the pavement while the façade is cleaned using waterborne abrasive.

With a couple of buckets of clean water and a sprayer, spray the surface until very moist (this will be absorbed by dirt and cause it to swell). Then get a nylon scrubbing brush wet and scrub in a regular pattern; use a clean moist sponge to lift off the grubby water and squeeze into a waste bucket, then rinse the sponge and repeat. Only apply clean water to the face, continuously rinsing brushes in a clean bucket of water reserved for this job. Keep on top of the task and always replenish with clean water. Keep dirt in suspension and give the wall a light mist rinse to get it off the surface. Try to keep the water runoff to a minimum; if it is going to be an issue get in charge of the situation and install a system to assist this.

Soaking and Draining

Most cleaning on a large scale with water involved, or where there is a need to wash off other materials used, will result in floods or at least puddles at the base of the building. If the area being cleaned, a window say, is small then it should be possible to tape polythene sheet below and have it lap into a temporary gutter with downpipe that leads to a drain, soakaway or water butt. Larger areas need channels of sheeting formed at the base on batten frames and sealed to the building, perhaps by taking out a low-level bedding joint before tucking and wedging the sheet into this.

The debris can be trapped by filtering through loose-weave cloth (fibreglass matting is available in many grades and is more effective than hessian, which lets dust through and is horrible to work with). Or you may be able to make a sump in a water butt where the water inlet is low inside and the outlet is near the top; increase filtering by placing a bucket full of easily cleaned or replaced clean gravel in the butt and inserting the inlet pipe into this. Place filter pads of layers of matting over any public drains nearby; keep them in place with strips of lead around the perimeter.

Nebulous Spray

For scurf removal there is more than one option. The gentlest method is to soften the material and remove it mechanically (brushing). Let us imagine a test area; the scurf is hard but technically water-soluble. A problem is that the surface is coated with soot and exhaust particulates (this gives it that toffee-to-black colour) which, being organically derived, can repel water, especially in big drops. To get the water in this surface, perhaps it should be removed first by mechanical or chemical means; for the gentler approach, reduce the water droplets' size so they are absorbed into the surface. This is one of the issues with water as the surface tension (meniscus) holds the liquid together preventing its absorption, up to the point where there is a lot of water being used without result. Try this out by sprinkling some water onto dusty stone and see how the droplets sit on the surface, then mist spray and see the difference. The trick is to create a very fine (nebulous)

Interior stonework in a Bournemouth church undergoing cleaning. A survey recommended the full cornucopia of peel-off films and poultice to do this, whereas a ten-minute trial discovered that a bucket of hot water and a sponge (with the occasional squirt of non-ionic soap, which acts as a surfactant and gets the water into the greasier stuff) would do the job perfectly well. Often people feel they have to justify their specialist position by making this work esoteric, when in reality much of it is basic and just needs common sense. When a technique is basic, try to get the people who live with the building involved in the work; they will appreciate it.

mist and apply this until the scurf is thoroughly wetted and softened, making it easier to brush off.

For a concerted effort rigs can be hired to mount on the face and supply a mist over a large area; to do it yourself, create a simple frame from sections of hose to face the area and have mist sprayer fittings at the junctions spaced so they cover the area required, fed by turning on a tap. This may seem to be getting technical but is surprisingly basic at one level; or it can become a serious bit of kit covering whole façades, using solenoid switches, computer-controlled by timers or by monitoring the humidity of the environment and stone.

For the artisan there is a quick fix using kit already mentioned; a compressed air spray gun (very cheap to buy) can be used repeatedly over the working day. For a clever solution have the compressor plugged into a timer, rig the sprayer so it is open, has a good supply of water and faces the area, and set the times for operation throughout the night; design and develop improvements as necessary.

Pressure Washing

There are specialist pressure washers designed for heritage cleaning that can be hired from the suppliers. Due to the requirements of grant-aided work, training is compulsory to use these; luckily the suppliers will do this on request as it means you will take more care of their machines and most likely be a return customer.

General pressure washers can be hired everywhere and if this will do the job effectively and safely consider them. Their use is simple; work from the top down, and keep the pressure as low as possible to minimize damage and the amount of water entering the fabric. Pressurized steam is far superior as the heated water is much more effective in softening dirt while the pressure removes it; they are more complicated to keep running but the amount of water is reduced and

Removing paint from a Coadestone garland and escutcheon, using a pressurized steam washer. This is extremely effective on hard substrates with oil- and acrylic-based paints. Note all the debris being generated which needs to be contained. Tests for lead content should be made when removing paints, as its presence will entail a completely different approach to the work.

because it is so hot a lot will evaporate off rather than soaking into the fabric. They are especially good at getting flexible masonry paint and oily stains off, as well as the usual surface grime that generally tends to be plant-based.

A decent water supply and power are going to be necessary, and also low-maintenance runoff systems are essential – as you will be too busy to notice what is happening away from the workface.

Pure Steam

Applying steam without pressure to clean stone is efficient, though will take time, as the nozzle area will be rather small to be effective; this is best used on delicate pieces and is great for marble. The trick here is to be very methodical, as with a lax approach the surface can end up mottled. On flat surfaces (this is good for inscriptions on plaques and tombs) have a strip of tape to the side and mark it as the level is reached. For less intense but larger areas, a useful variant is to soften up dirt and scurf by placing the plastic hood of a wallpaper steamer over the area until it is ready; although it is not as hot it does work, and there are industrial models that improve speed and effectiveness.

Using a small steam jet to clean up a lettered plaque. Really only suitable for the finer jobs, it is quite slow and large volumes of material will cool the vapour.

Poulticing

The concept of poulticing is to apply and hold whichever chemical (including water) is being used to clean with, holding it against the surface of the stone in a thixatropic paste; conversely it can be used to soak up liquid into which unwanted materials have been pulled, so they can be disposed of.

The active ingredients that are going to do the job are usually made up in solution, which is mixed and held in an inert medium, usually very fine clay such as attapulgite (a type of fuller's earth), sepiolite, paper pulp or cellulose.

Mixing

Wear protective disposable gloves, eye protection and a suitable mask; overalls are highly recommended. Take photographs of the surface before, during and after poulticing to track changes in appearance.

Dry materials should be mixed together first, usually by weight, so a kitchen scale is needed. Remember to put the empty container on the scale and tare it first, then weigh the ingredients and put them all in a clean mixing tub and stir with a wooden spatula. The liquids are then mixed together, usually by volume, so use measuring containers here; if solvents are involved use glass beakers or at least something

Practical poultice application using a plastering gun and hopper. This gets the job done faster, more cheaply and more effectively than working by hand when cleaning large areas.

Localized poulticing applied to remove copper staining caused by run-off from the bronze statue above. This situation is a maintenance issue now, as the accumulation of metal salts in the stone means staining recurs on a regular basis.

that will not melt. Put the liquid into a large tub and mix thoroughly with a drill-mounted plaster-mixing paddle. Add the dry ingredients in a steady stream to prevent dusting up. Leave the mix to stand so that it becomes a sloppy mass, thickening up as the medium absorbs the liquid.

Application

Wet the surface of the stone by spray and, with a gauging trowel or spatulas for the nooks, apply the poultice in an even layer, ensuring complete contact is made with the surface. For large-scale applications a compressed air plastering gun (as used in general construction) with a top-loading hopper will cover large areas in minutes with an even coat. As an added bonus the force of application ensures good contact with the surface. Remember to mask off non-treatment areas with polythene sheeting and gaffer tape to reduce overspray. The nature of a medium such as cellulose makes it difficult to layer up vertically; one way to help is to put sheets of kitchen roll on after application to reinforce it, or possibly put chopped strand in the mix.

Protection of the poultice to prevent drying out so that the chemicals can do their work on the stains which, once 'loose', can move into the poultice and be lifted off.

To Dry or Not to Dry

Some poultices need to be left uncovered to dry out quickly; others will need sheeting over to prevent this and keep the ingredients in solution, to be removed when still damp; and some will need to be covered over while reactions are taking

Here the soaking and removal properties of the poultice can be seen to be doing their job by the colour of the salts in the material.

place and then uncovered to dry out before removing. These are all dependent on the role of the poultice and how long it must be left on. To keep track of everything, stick tape on the wall and write the time of application, references and how long it should remain; this is the dwell time and should always be noted down as well as any comments on the process – strange things can happen occasionally, so good recordkeeping is paramount. Lift the used poultice off with wooden scrapers (do not use oak as it contains tannic acid and reacts badly with lime) and have sturdy plastic sacks or tubs ready to take the used poultice off site for disposal. Never reuse a poultice.

Air Abrasive

Not so long ago, the cleaning of buildings was limited to sand-blasting by individuals wearing suits looking for all the world like astronauts in a 1950s science fiction film. This method used high pressure, and copious amounts of sharp hard grit would tear the surface of the stone off, destroying detail and leaving a devastated surface, ready to get soiled straight away; perhaps it was a highly advantageous built-in obsolescence and job creation scheme for us later generations? In these enlightened times we now use much gentler abrasive systems shot at an angle, calming the violence by adding water to the mix. 'Swirling vortex' is a name given here, though there are many variations on this theme; the idea is the abrasive knocks off the dirt rather than eating it away. The type of aggregate is an important factor here as it can still erode the stone if too hard, so check out the range of available abrasives and choose accordingly. Hiring or buying this equipment is a personal business choice, though even with a medium project the cost of buying the kit can be recouped quite rapidly; the downside here is that spares, updates and maintenance all need paying for rather than being the hire company's burden. The same applies to the pressure steam cleaning kit; this can be hired and training acquired at the same time so the outlay will be recovered quite quickly and useful knowledge gained at the same time.

Using the Jos system of swirling vortex air abrasive to clean complicated stonework. With a good range of nozzle sizes available and controllable pressure these systems can work on massive buildings right down to very fine and delicate material. Traditional micro-sandblasting using aluminium oxide is pricey and unhealthy, and in the wrong hands can do damage – but that is really true for most of this work.

Tips

Do not wander all over the surface, no matter how much fun it seems; be controlled and aim to do one stone at a time. This will ensure evenness of clean, and on lightly stained buildings it will help keep track of what is done; this sounds odd but as the surface around the cleaning area gets wet the stone darkens and you can be misled about which area has been finished.

The uninitiated will more often than not turn up with basic protective kit, usually a site helmet with visor and wellington

Site-designed visor wash, using a converted safety helmet and some plumbing kit, is fuelled by a side take-off from the nozzle water supply, so that when you stop working the washer does as well.

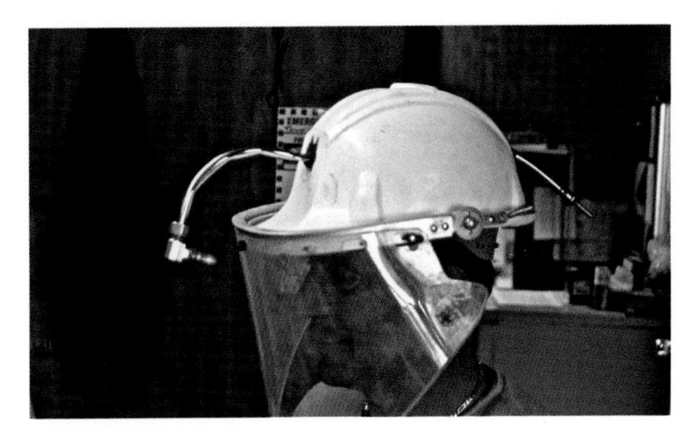

Open air cleaning in places such as here in Bunhill Burial Fields has its issues due to public presence and containing the mess. Sheeting was at the base and much of the calcium carbonate dust which is relatively harmless dissipated into the grounds.

Cleaned Bathstone plaque had light sulphation and grime removed using a clay poultice and 5 per cent solution of ammonium bicarbonate left on for twenty-four hours; this is a good chemical for greasy stains and softening of scurf, and is used in many proprietary mixes.

boots, so by lunchtime they are soaking wet, have grit in their eyes and are temporarily deaf from the compressor, wanting to go home and definitely no good for an afternoon session. For serious work and comfort, full waterproofs, with rubber gauntlets over the sleeves, and a forced air helmet (preferably with a visor wash installed) will make life in the maelstrom of mist, grit and noise almost bearable.

With an indelible pen, note on the machine the pressure that gets the best results and at the end of the session/day note how much has been cleaned, the amount of abrasive used and any observations, and check the area next day before commencing work to see how it has fared when dry.

The control of runoff is as for water washing, but with lots more dust. If working near other buildings or vehicles be aware that the dust-laden vapour generated can travel quite far, so have a hose ready to rinse off surfaces and cars before people get bothered. Keep all the abrasive material bone dry, storing it in covered containers well away from the workface, as it will clog up the pipes and valves with even the slightest moistness. Do not reuse abrasive even if it looks really clean.

Chemicals

Using strange chemical mixtures on a stone is fraught with issues. Obviously there has been research into and experience of how they react with the materials they must attack and the materials they must leave alone, but there will still be surprises – so know your materials. The only maxim here is that even though a situation calls for this type of intervention, always carry out a trial first – just in case.

There are two types of approach here. One is where the chemicals are applied directly onto the stone, by spraying or brushing, and the other is to hold them in a poultice, which is probably the best way to keep control. When using a poultice with chemicals in, always have the material safety data sheet (MSDS) or relevant Control of Substances Hazardous to Health (CoSHH) information; to comply with health and safety rules this information should be to hand for every material used.

Have neutralizing agents available and soak-up material (poultice clay is good for this) for spills and be aware of any medical needs should there be contamination or problems. Know the regulations for disposal of used material, and stick to them – most of these are environmentally hideous and should not end up as landfill.

Brushing and spraying has the same implications, though be aware of dwell time and cleaning off, so do not do massive areas in one hit, hoping to speed the job up – this could turn into the 'Sorcerer's Apprentice' moment. If there is exposed metal in the area to be cleaned protect it with a coat of quick-drying varnish, shellac or similar, as some acids can be corrosive to metals.

Chemical cleaning trial in Wakefield Cathedral brought the stone up really clean but left the pointing dirty, creating an unsightly contrast. The position of the panel in the middle of a wall pays little heed to the notion of carrying out trials in a discreet area.

Off the Shelf

There are numerous materials manufactured for the cleaning of stone; entrepreneurs will always exploit a good market and heritage work is one of the major growth areas in the construction industry. This is probably a good thing as it makes specialist materials, once scarce or untested in the field, available by next day delivery in some cases and certainly speeds things up. The issue here is the one held in highest regard by all those professionally involved in this work (if it were not true there would be no need for us conservators): that all buildings/materials have unique properties and issues so any intervention/treatment should be tailored to such by qualified and experienced practitioners. Put simply: if you had an extra arm, a suit off the rack would be useless, or if bought would need extensive work to fit; so what is the use of universal treatments for singular problems? If a treatment must be bought off the shelf, it is imperative to know exactly what is in it and what it was developed on; if possible speak to the supplier and discuss the problems. If they do not ask what type of stone or dirt is being removed or give bland general statements, then perhaps it may be prudent to consider alternatives.

Brush-on film trials, in a low out-of-sight corner, using two different strengths of cleaning agent.

Peel-Off Films

Some treatments, though classed as chemical, are not especially exotic, though the suppliers would have us believe this; there are many variations nowadays of brush-on latex film laid on and peeled off taking lots of grubbiness with it. Be aware that once it has a proprietary name stamped on the tin, perhaps with the addition of a few lauded, sexy-sounding additives, the cost rises exponentially.

While not as extreme as a chemical wash, the trials have cleaned the mortar as well; so this is a softer but more practical option.

The basic process here is that a rubber (latex) is worked onto the surface and after sticking to the dirt particles does not let go of them; this film is then peeled off leaving all clean. Some add chemicals such as ammonia, a caustic and hazardous ingredient in many cleaning products, and EDTA (ethylenediaminetetraacetic acid) to help the process get more dirt off.

Make a Comparison

As an exercise get hold of some samples from various suppliers. Find a dirty bit of stone and mask off around a number of reasonable patches. Apply the film as per instructions to some and on others apply a film of some good-quality silicone rubber (food grade is best, but this is a practice, so use what you can get) and also try liquid latex. On one patch mist-spray with hot water, on another apply a poultice of hot water in a medium (mashed-up hand towels will do at a pinch) and leave for an hour. Give all these a good wash with brush and sponge.

When cured remove all the films, neutralize the surfaces if the manufactured ones demand, and then compare all the patches; I guarantee there will be some interesting results!

Cement Removal

Occasionally stone has been coated with OPC by some bright spark hoping to stave off the loss of the stone or freshen up a decaying surface, thus hastening the decay. The only thing that will dissolve or soften this is acid (usually hydrochloric); unfortunately it is also pretty lethal for a lot of stones as well and incredibly hazardous to use in quantity. Try the softer viable alternatives such as mechanical removal (a technical term for using force) first if possible.

Cement applied to stone usually exists in one of two states. It is either completely adhered, forming a tenacious bond that will take some surface material with it when getting it off. The other is that the cement is not stuck at all; there is a definite gap between it and the stone, but it is held in place by being stuck elsewhere through its own physical rigidity. With the first state, realize it is going to be a job to get it off and price accordingly (based on the worst case scenario).

Chiselling off cement from elements like this Coadestone requires sharp chisels and precise cutting to avoid damaging the original material; get some practice on this until techniques are mastered.

Splitting Up

The thickness of the cement is important so if this is more than 4–5mm, it may be necessary to wear it down. This is possible using micro air abrasive tools or careful grinding and working away at it carefully without touching the stone. The

A church doorway in Cornwall was of a micro-granite extensively coated in cement wash. The removal of the well-attached areas was impractical, so only loose material was removed and a mortar was used to give a unified appearance.

issue with this is that if it is aggressive enough to wear down cement, it will certainly erode stone so great care must be taken. For isolated lumps of cement or foreign material stuck to stone, mask off the precious area with gaffer tape cut to fit exactly and then spread silicone sealant or other rubbery mastic on the tape; when it is dry the abrasive will bounce off this, and concentrate its power on the blemish.

For thin layers of cement, get a very sharp TCT chisel, 12mm or less in width, and hold this at right angles to the surface with the blade resting on the cement. Grip the chisel firmly and strike it with a sharp hammer blow; a scallop of cement should pluck off. The trick is to prevent the blade going through to the stone, the impact aiming to fracture rather than cut. Start close to the edge and 'bite off' small pieces in lines; with practice this can be a smooth, effective operation.

If the cement is laminated away from the stone, it should be cracked into smaller pieces; with good judgement the layer will come off in plates without damage. Be aware it may be adhered on mitres and so violent pulling off here may lose arrises. Pitch with a sharp TCT chisel across the flat of the surface (such as a fillet) to create an edge to the cement, which can then be pitched the other way off the stone.

This room had been plastered with a porous mortar to promote evaporation of moisture from the perpetually damp walls. Unfortunately the owner purchased a limewash promoted as correct for this (that is, that the walls could 'breathe'), but was actually made with a large amount of acrylic and thus sealed the wall, creating more problems. Rather than replaster the whole room, it was decided that the surface could be cleaned of this paint using a dry abrasive (the waterborne method was not an option here) on low pressure that did not cause damage to the surface.

What is Used and What It Does

It will help when starting a cleaning project to become familiar with the types of materials used and have an idea what they are supposed to do. In itself this section could easily become another library of material science. There is a lot of information out there on the exact workings and processes so dig deeper. This is just an aide-memoire to give an idea of how they are chosen; it is down to the results of trials to see if they should be.

Abrasive Aggregate

- Aluminium oxide; used for finer work, such as removing scurf from carved detail.
- Glass microspheres; used for gentler work where less abrasion is needed; this can be termed peening.
- Calcium carbonate; basically limestone dust, its hardness depending on the original material; also includes marble dust.
- Magnesium carbonate; Dolomitic limestone dust.
- Bicarbonate of soda; used for gentle cleaning of polished materials.
- Walnut shells; used for delicate soft work.

Acidic Cleaners

These have pH values of 5 and lower; those with lower values are stronger and are good for removing mineral and metallic soiling.

- Acetic acid (ethanoic acid), CH_3COOH. Vinegar is 4–10 per cent acetic acid.
- EDTA (ethylenediaminetetraacetic acid). Colourless, water-soluble solid. Widely used to dissolve limescale, it is a chelating agent (grabs metal cations and pulls them out when in solution).
- Hydrochloric acid, HCl. Clear, colourless, highly pungent, highly corrosive mineral acid (found naturally in gastric acid).

- Hydrofluoric acid, HF. Colourless, acutely poisonous and highly corrosive; will eat through glass. Was used widely for cleaning sandstone in the past.
- Oxalic acid. Crystalline solid dissolved in water; used as a reducing agent (to arrest oxidation of iron) also as a chelating agent.
- Phosphoric acid, H_3PO_4. Used as a rust inhibitor.
- Sulphamic acid, H_3NSO_3. Similar properties to hydrochloric acid but less corrosive.

Alkaline Cleaners

These are good for altering the pH in order to get ions mobile; they are used for greasy and biological soiling, oil-based paints and breaking down scurf.

- Ammonium, NH_4^+. This gives off ammonia, which is caustic, hazardous and stinks of urine.
- Ammonium bicarbonate, $(NH_4)HCO_3$. White solid, good for cleaning greasy statuary and scurf.
- (Sodium) hypochlorite. This is the main ingredient for most bleach.
- Potassium hydroxide, KOH. Caustic potash, similar to lye.
- Sodium bicarbonate, $NaHCO_3$. Baking powder.
- Sodium hydroxide, NaOH. Caustic soda or lye. Caustic, used for making soap (see the movie *Fight Club* for a good perspective on this).
- Dichloromethane or methylene chloride, CH_2Cl_2. Solvent, widely available as a paint stripper.

Organic Solvents

This means petroleum- or plastic-based materials; these can be known as aromatic hydrocarbons and are solvents of a wide range of materials.

- Acetone, $(CH_3)_2CO$. Also called propanone. Colourless solvent, miscible with water; simple ketone.
- Ethyl benzene, $C_6H_5CH_2CH_3$. Toxic, highly flammable, colourless liquid that smells like petrol.
- Toluene. Clear, water-insoluble liquid that smells of paint thinners; it is a toxic solvent (but has been used to extract cocaine for Coca-cola!).
- White spirit. Also known as turpentine substitute.

Surfactants

These break down the surface tension of mixes and thus aid mixing.

- Ethanol. CH_3CH_2OH. Ethyl alcohol, pure alcohol, grain alcohol, or drinking alcohol. A volatile, flammable, colourless liquid.
- Glycerol. Polyol (or sugar) alcohol, which is viscous, colourless and odourless.
- Non-ionic soap. A widely available non-reactive detergent.

Pulp Medium

- Attapulgite, sepiolite, fuller's earth; these are finely ground clays, commonly used as wallpaper paste.
- Carboxy methyl cellulose, CMC, commonly used as wallpaper paste.
- Cellulose or paper pulp.
- Acid-free tissue or paper.

Queen Victoria's statue, Weymouth, was cleaned using pressure washers and some localized poulticing; eroded and missing stone was fixed using mortar repairs and Dutchmen. The bronze was causing verdigris staining on the plinth, and the metal was quite dull so a microcrystalline wax (softened using white spirit solvent) was applied to invigorate and seal the metal; buying this wax in large lumps from casting suppliers is much cheaper than small tubs of ready-made product, and softened with white spirit it makes a good protective coating for metals, alabaster and marble.

CASE STUDY: LEAD FLASHING

Many stones do not really fare well if they are exposed to the elements in positions such as copings or projections; others have lovely carving that cannot cope with a lot of elemental wear. A good practice is to give an added layer of protection by covering them with lead to alleviate the worst of the weather.

Cills on the town hall restoration in Aix-en-Provence. There is no weathering (sloped) surface to these so water could sit on these and get up to mischief.

A desultory flap of lead hangs over a delicate carving on Sherborne Abbey, a lacklustre approach that would have been improved 100 per cent if a simple gutter had been put in by turning up the bottom edge.

A bit more complicated on our project in Winnipeg, the lead was to prevent standing snow (often two metres deep) melting into the stone.

Beating the lead flange into a reglet. It is essential that lead is fixed securely as it can lift in strong wind and flap about or come adrift. On large sheets, allowance must be made for the thermal expansion which can tear the metal off its fixings; there is information easily available for the use of this material and its performance – handy to have.

Protective covering to Coadestone shield. Lodged in the wall, it prevents water running down the back of the piece where debris can accumulate, get trapped and let moisture seep through the brickwork into the timber just behind.

347

CHAPTER ELEVEN

CASTING

A viable alternative to making repairs to masonry and carved detail with stone (and other materials) is replication, by creating missing elements using casts. To be able to do this well is an essential part of the restorer's repertoire and will have many applications. Benefits include being able to make copies of personal work to sell or use as promotional material, and keeping examples of work for reference and interest. We will not go into the ethics of whether using replicas to replace parts of a historical structure is acceptable, although it is worth mentioning that mortar repair is just that. Also, for repeating elements that have aesthetic value (balusters, modillions and others), then providing good-quality and economical replacements can mean money being available for more deprived areas.

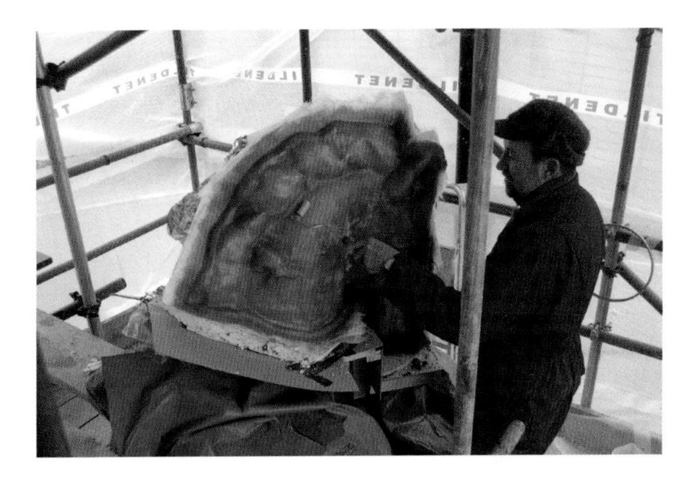

***In situ* mould making. The existing head of the unicorn was made in *cimente-fondue* attached to a Coadestone statue; the head has rusting iron armatures inside and will eventually be lost when it finally breaks apart. The new cast was made and stored away to be ready when this happens.**

Main Types of Casting

The two types of mould (here this means the container out of which a replica is produced) most used for this work are case moulds made in the workshop and squeeze moulds made on site. These two general descriptions will cover a huge variety of applications that in my personal experience include small fleurons and fifteenth-century angels in fake wood, balusters in 'Bathstone', a 3 × 2-metre white 'marble' fire surround, and an imperial eagle in Roman cement, amongst others; so follow the instructions, get some experience and start developing.

OPPOSITE PAGE:
Replacement eagle for the façade of the Egyptian House, Penzance. The original had lost all detail and repairs had turned it into a travesty of design. I was commissioned to design a maquette based on old photographs; this was then modelled full-size in clay and cast in Roman cement with LECA aggregate.

Material Matters

The materials used here are all readily available, with basically two ways of getting hold of them: going down the artist's or hobbyist's route means buying small amounts at inflated prices. These suppliers tend to have a limited range of stock, relying on a 'one size fits many' approach to mould making. More sensibly, purchase from industrial suppliers, where buying larger quantities significantly reduces the cost and gives plenty of spare material for playing with later; I tend to price casting jobs based on the expensive costs of small amounts, and buy from industrial suppliers, which reduces profit on one job but increases profit in the long term.

Ingredients

We will skip the science and just cover the general applications and traits of what is going to be used; experience and practice will build up confidence in getting the best out of them.

GRP

Glass reinforced plastic, or fibreglass, consists of sheets (mats) of hair-fine glass fibres set in a resin; this is strong, light, durable and, if needed, flexible.

Mat

The mat comes in weight by square metre, for our purposes the heaviest, 600g/m^2, is the choice, as the lighter ones just need more layers. It can be bought by the metre but is best to get a large 25-metre roll that will last for a long time.

Chopped strand is the economical version and good enough for this work; it also comes as woven glass fabric (stronger and dearer), bandage or roving for wrapping work or edging (where it upgrades flange strength considerably), and surface tissue to give a clean smooth finish to the outside of cases. Loose chopped strand is good for reinforcing casts internally (in plaster and resin where there are delicate projections), but rather than buying this, just cut and tear up some mat if needed.

Resin

Resins used in GRP are either epoxy (rather specialist, expensive and not used here) or polyester, which is the choice for the general boat-builder, flat-roofer and us. Buy general-purpose resin in 20–25kg tubs for economy; the catalyst will be included.

Benefits of having GRP kit around includes being able to fix dowels in stone, make repairs to cold-cast bronze or iron and, importantly, carry out repairs to cars, boats, surfboards (though not high-quality repairs), roofs and many other items.

Rubber

The range of silicone rubbers (or elastomers) available for making moulds is pretty big, covering all aspects of performance (and price) and impossible to cover here, so when choosing, read the literature, talk with the suppliers or try contacting people like me who are always up for a friendly chat, to get the right rubber for the job.

It comes as a two-part mix, with various catalysts that can speed up or slow down cure time; also there is thixatropic additive available for some that can make it thick and putty-like for paste-on applications. Builders' merchants carry a range of silicone mastics, that cure without catalyst, for various purposes; choose with a use in mind and read the label to know what is actually in the tube.

DON'T DO IT

It may be that cost makes the option of latex or polyurethane rubbers attractive; resist using these anyway. I do not use them and suggest that you do not either – you will thank me in the end.

Solvent

Acetone is the solvent and general cleaner of choice; buy a five-litre container and store in a safe place, as it is extremely volatile. Have a sealed tub with some in the bottom into which brushes and paddle rollers can be dropped when work is finished. This will help to remove resin effectively; if this is neglected then the rollers will seize up and be useless, while brushes will set solid.

Sundries

Have a good supply of disposable brushes for applying resin, also disposable gloves and paper towel to mop up spills.

Poured Mould

For this exercise the object to be cast is a baluster, the original from which the mould is taken being called the master. Variations for other shapes will be mentioned in passing; the pictures accompanying this will give another project.

Set up a sturdy workbench with a flat top and staple a polythene sheet across the top to prevent resin drips; when finished the whole mess is rolled up for disposal. Get a smooth-faced board about 60cm square; melamine-faced MDF or similar offcuts are best as the board will be useless afterwards. This will be the baseboard, and is placed on the bench smooth side up.

At a bedding end of the baluster where the piece is fixed, run a silicone bead just in from the edges and firmly place this end down in the centre of the baseboard; leave to cure. Cut a piece of 18mm plywood (have this prepared, waxed or varnished already) the same shape as the exposed end of the baluster fixing bed (the top for this exercise), with an extra 50–60mm over all the way round and silicone this to the top of the master. When it is fixed tightly, completely cover the master with aluminium baking foil, secured with masking tape, to prevent it getting dirty from the clay.

Clay Board

For getting consistently regular sheets of clay, make a rolling-out board by pinning two 6–8mm strips of wood just under the length of a rolling pin apart onto a smoothed piece of plywood. On this place lumps of clay and roll out with the rolling pin to make a sheet of even thickness. Fresh clay can be messy to work with; to improve this leave the bag open for a couple of days to stiffen up and before use, pound it soundly in the bag with a mallet or the rolling pin.

The Blank

This stage is to form the shape of the rubber part of the mould in clay, over which the outer case (made of GRP) can be made. The case is as important as the rubber, as it holds the mould in place correctly, so invest time and thought into this process – bad case, bad cast.

Clay Work

With putty or stripping knife, section the sheet up into tiles that can be lifted off without tearing. Using these tiles, making more and shaping as needed, tightly encase the whole baluster in an even layer of clay. Do not pull the tiles thin over the edges; instead fold or butt them up. The aim is to have the same thickness all over – thin areas may weaken the mould or ruin the cast.

At the base lay tiles around on the board to make a 30mm-wide flange off the baluster, which when made in rubber will be the lip to the entrance for the casting material and lock into the case. Trim this with a slight chamfer and round the corners.

This is going to be a one-piece rubber mould, but to get the cast out it will need to be slit from top to bottom. The location of the cut will be down to common sense; it needs to be manageable and discreet, along corners or ridges for example. It is possible to make sleeve moulds that can be pulled off the master and cast without cutting, but this depends on the shape and delicacy of detail; if there are deep undercuts or projections the fight (and it can be a struggle) to get it off and the pressure from the rubber can cause breaks. Objects with a large back or base, acanthus brackets for example, can cope with whole rubber moulds.

The slit will need to be locked together in the case so a raised flange of clay (rubber eventually) along the line of the proposed cut, about 10mm wide and deep, is formed from the top to the bottom on the clay, to lodge in a recess in the GRP.

Easy Pull

When adding flanges and projections, always put a radius on internal mitres, then look at the blank and consider that as the GRP sections will be quite rigid, they must pull off without catching; build up with added clay to install gentle slopes and fill out depressions to abet this. Do not pull the clay tiles about as this will thin the rubber and possibly lead to tears.

Inlet Tube

When pouring rubber it is best practice to fill from the bottom to prevent air being trapped, causing collapsible weak pockets in the mould.

Lincoln Dragon. A replacement decorative roof grotesque had been carved by Michael Thacker for Lincoln Cathedral and it was decided to create a limited run of cast copies to raise money for restoration work to the church.

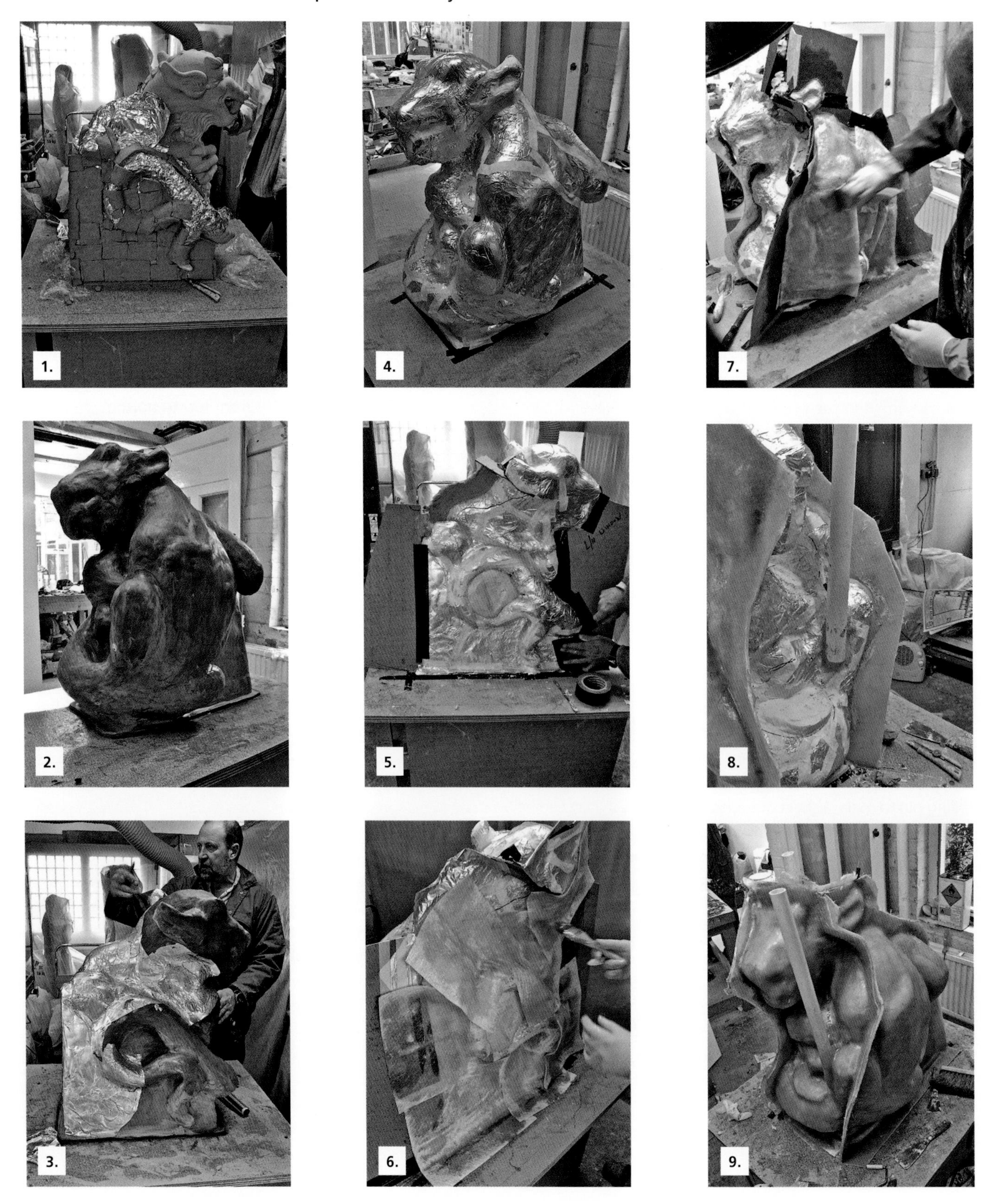

1. Covered in foil, the carving is covered in clay tiles to build up the shape of the rubber.

2. The clay is smoothed over, without thinning it out.

3. Covering the clay in foil to isolate it from the GRP.

4. Edges of foil taped down and smoothed ready to start the laying-up process.

5. Blanking wall for building up flanges is erected around the first section area. The pale plug in the centre of the foil will be a detachable piece to simplify the shape of the mould and give ease of dismantling.

6. Sheets of fibreglass mat are resined up and placed on the blank; they are not padded in until the resin has spread and softened each sheet.

7. This is a multi-section case and here the next section is walled off and laid up, leaving space for fill-in sections and a top piece.

8. The fill tube is mounted on the body of the blank so that the rubber will flow from the base up.

9. Completely glassed up with fill tube and air release tube enclosed in the GRP. Holes will be drilled, the assembly dismantled, the clay removed, and then reassembled for rubber pouring.

10. After the rubber has cured the sections reveal the rubber; note how the rubber has crept into the flanges, forming a flappy edge.

Install a pour inlet made from (ideally 32mm) plastic plumbing tube running up next to the mould, but not touching (it will be cut off when demoulding the master), and finishing a bit higher. Stick this to the baseboard about 50mm from the flange with silicone and let it cure; it can be steadied at the top with some wire stapled to the topboard. At the base form a runnel in clay, so that as the rubber is poured down it feeds into the base of the setup.

Isolation

Polyester resin will not cure well on the damp surface of clay, so this needs isolating or coating. PVA will do this, or use aluminium foil stuck to the clay and cover joints with masking tape; get this as smooth as possible as this will be the outer surface of the rubber.

Outer Case

When designing a case for moulding rubber, it must be capable of being disassembled and lift off the mould easily without retention from undercut or protrusions. In the case of a baluster a two-piece case with a base plate will suffice; the joints will run vertically on the corners, and the cutting flange is in the middle of a section (this prevents distortion when bolting up the case).

For complicated casts more sections may be needed; look at the master and figure out the shape of panels that will cover it but also pull off easily – to get the effect imagine an exploded technical drawing of this.

Flange

The case sections will ultimately be joined together by flanges, and made one piece at a time, so a wall for the flange to be built against needs to be added to the foil-covered clay now. This should be strong and secure as it takes some pressure whilst laying up the GRP. Use two hardboard strips that fit between the two boards and have the roughly cut contours of the baluster section where the flange will be and projecting out at least 75mm. Wedge these in place with a smooth concave mortar haunch for a surface that will eventually be the finished GRP surface. To aid assembly of sections stick a couple of rounded clay bosses on this flange, cover with foil (taped around the edges) and smooth down; the result will be nubs that lock into recesses on the joints – handy for long panel flanges, when there is a chance of misalignment. The haunching should flow smoothly onto the clay blank. Brush this (it is difficult to get foil on), the whole of the blank and boards on the side and the runnel piece (but not the body of the inlet tube as the resin must stick to this) with a mix of petroleum jelly and white spirit (10–20 per cent) as a release agent; there will be small niches in the foil that can be levelled out by using a brush to smooth out the lumps of jelly lodging inside.

Laying Up

This is the smelly part, so the correct breathing mask is required, also lots of disposable gloves, overalls and old boots (resin drips on leather ruin the finish).

Making GRP is not complicated and does not need to be (too) messy but it needs a steady, calm regime; there is a time limit on the gelling of the resin, but work without fuss, panic or rush and it all comes right.

Ready to Start

Assemble all that is needed before starting: disposable brushes, mixing tub and mixer (stick), knife and new blades and paddle rollers. Prepare enough mat ($600g/m^2$) to easily cover the side of the blank and flanges in three layers; cut six large pieces so the side can be covered in two pieces that overlap well, and a pile of 6 × 30cm strips for flanges. Use hard ground scissors for cutting smallish pieces or a sharp Stanley knife on a cutting board to slice up larger sheets. The knife blade must be sharp as the mat will blunt blades, which then produces a messy torn edge.

Pour the resin into a plastic mixing tub, about half a litre to start with, and add about 3 per cent catalyst. Mix well.

Applying Mat

Take one of the larger pieces of mat and spread it with a brushful of resin; place this, resined side in, on the bottom half of the blank and push into place with the brush. Do not worry that it does not go in neatly – the resin will spread through the fibres and soften the mat in a minute. Resin the next and cover the rest of the blank, then pad the first into the blank to fit closely.

Resin up flange pieces and place them along the flange, overlapping the ends and sides to the main body; pad them in after they soften up while laying the next one. Working methodically, lay three layers on the flange and blank. It will be tricky doing the ledge of the topboard so reduce the flange width to help it stay. Build around the base of the inlet tube, trying not to trap air by bunching up the mat, and blend the fibres onto the tube – this seal is important.

Once covered, take the paddle roller and work across the surface out to the edges of the flange, to push the resin throughout the mat. Trapped air, which will show as grey splodges, needs to be gently forced out through the edges; be aware of the wet mass's tendency to pull out of corners while rolling so press into these with the roller as needed.

All the above should have calmly taken place within the gelling time (about twenty minutes). Once the resin in the tub starts stiffening it will not spread, so stop using it. If the resin runs out, do not mix up a new batch; finish the above even if there are not enough layers on, and when it is cured add the missing layers as above. Once some experience is gained it may be possible to mix more during process, but it needs to be under control.

Trimming

Keep an eye on it now and, after a couple of minutes, with a very sharp knife press into the edge of a flange; when the blade slices in without dragging fibres it is 'green' and ready to trim. Cut the excess material from the flange to give a neat edge right around the case section, stopping if it starts to drag and tear until it stiffens up. The cure time will depend on temperature, quality of resin and amount of catalyst.

This is important to do well, as once the GRP is cured the only way to trim is with disc or modified jigsaw blades, which produces hideously itchy and dangerous dust and is not fun or easy.

Next Side

Once the GRP is fully cured, remove the walls for flange building and clean any detritus (clay, foil and tape) off the GRP surface. Where the flange meets the blank will need filling with a bead of clay, haunched to form a nice junction for the sections; be careful as any loose fibres sticking out here will be resined and like needles; resined fibre splinters are extremely painful and, being clear, are really difficult to find and remove.

Fold foil over the edge of the flange to prevent resin from the next section running over and sticking to it, then brush all over with the release mix and carry out the same laying up as before, trimming the edge to the existing flange.

Preparation for Pouring

When all is cured, drill 8mm holes through the joined flanges away from the blank side for the case bolts. Position these near the boards and in the middle and also through the topboard and its flange (one on each side should suffice, although this depends on size). These should be placed asymmetrically so that they are easy to match up again later. The dust from this is bad so have a vacuum tube next to the drill area to clean as it happens.

On the baseboard flange part, drill an 8mm hole through each section to place a short bolt as a location marker.

Dismantling

To take the structure apart, wear thicker gloves to protect against glass needles and push two levers under the topboard (old woodworking chisels or scrapers are good for this), levering it off carefully – force will be needed.

In the seam between the GRP flanges, insert chisels and twist them to break the case apart. Be careful, as the master must not be knocked loose; slide off the sections and clean all the insides. The inlet tube should be attached to one of the sections where it was butted against the baseboard. Cut a hole to let the rubber enter the runnel.

Remove all the clay blank from the master and flanges, and give the master a light spray of release wax – do not clog the pores, as texture will be lost. Brush the release mix over the insides of the case up to the haunching on the flanges, and inside the inlet tube.

Sealing

Run beads of silicone along all the flanges so that when put together it will completely seal the inside of the case – this must be absolutely spot on as the rubber is incredibly insidious and leaks are the last thing needed; they will be costly and may entail doing it all again.

Reassemble the case around the master locating it on the baseboard with bolts in the holes, put 8mm bolts with butterfly nuts and washers on both sides through all the flange holes and tighten up. Around the base flange insert some self-tapping screws to hold it down; do not over-tighten these as the case could distort, only enough pressure to grip the silicone seal, which should be fully compressed in all places now. Use dark silicone and any missing areas can be spotted.

While it may be completely sealed, experience dictates that 'belt and braces' is the best approach, so now run and trowel in silicone beads along all the edges of the GRP as a secondary line of defence.

The topboard in some moulds can be left off as the rubber will level out and allows the flow to be better, or drill a large 20mm+ hole or two to let the air flow out and attach it to the case with bolts as above.

Let this cure.

The Pour

Silicone rubbers are quite expensive and so it is best to mix up enough to do the job without (too much) waste. To guesstimate the amount take the clay used to make the blank and pound it down into a tub, this should give a volume close to the amount required to fill the case; remember there is the inlet tube volume as well. The inlet tube volume will be wasted money, so in some cases the inlet can be incorporated in the blank within the case, designed so that the rubber will head straight to the base of the mould first then back up.

Make a funnel for the inlet tube by cutting a plastic water bottle in half, enlarging the mouth and siliconing it in place; if this is high, it will be awkward to pour into, so a hop-up may be needed. Do not put the assembly on the floor as peace of mind requires that it can be closely inspected during this stage.

Decant the amount of rubber needed into a clean tub. This is messy stuff so have a trowel or scraper to cut off the flow and flick up the overflow back into the container.

Measure out the catalyst as instructions, by volume or weighing on scales, into a disposable plastic beaker and pour into the rubber; mix slowly and thoroughly for a good ten minutes, using a wooden paddle, keeping it deep in the rubber to minimize air entrainment. If you can get use of a vacuum chamber this is really helpful; place it in and pull out the air. The rubber will rise and bubble so make sure the tub is bigger than the volume by a good amount; if not available let the mix stand to allow as much air to bubble out as possible. The gelling time is quite long with standard silicone mould rubbers so do not be hasty.

Now pour the rubber into the funnel slowly and consistently at a rate that does not let it start folding onto itself, as this will trap air. Halfway through, stop and give it another good mix; if silicone rubbers are not thoroughly mixed they will never cure and could ruin the mould.

The rubber will flow up the case pushing air ahead of it. If the shape of a case includes potential air traps, drill 6mm weep holes at the top of these and as the rubber reaches and flows out, stop up the hole with a lump of plasticine and strip of gaffer tape over to keep it in place – do not use clay as this will just pop off.

Once filled and there are no problems, let it cure.

Demoulding

Leave for at least a day to complete the cure. If using old rubber then it may take longer, but it will eventually be ready to get out and use.

Cut off the inlet tube at the base using a hacksaw for the GRP and a knife to part the rubber, remove and bin. Take all the bolts and screws out and lever off the topboard. This will be harder than before as the rubber is quite tenacious and it may be necessary to slide a blade onto the flanges to break the seal and a stripper knife to peel off the rubber as it proceeds. Wear gloves to protect against splinters, as before.

Split the case and lever off the rubber inside. This will involve spreading the sections, popping in wedges to hold it while releasing the rubber further in. Do not worry if small tears occur where the rubber has entered the GRP; the job is to get them off and leave the rubber-encased master.

If is to be a sleeve mould, now is the moment to slide a wide blade under the flange on the baseboard to loosen the rubber and then loosen the rubber from the master by resting the top against the chest and pulling up to break the suction. Then start moving the rubber up the master, with loads of jiggling and force; this is very physically demanding and will take much sweat-inducing effort.

Trimming the flappy edges to the raised haunches in the rubber. This is a one-piece rubber mould so there is no cutting for seams.

For the cut version a slit in the prepared flange must be made by lifting the rubber at the base away from the master and worming a flat blade in and levering, possibly using chocks to help here. Then, with a very sharp blade (a specially shaped blade for this can be bought – and is preferable), make an incision at the edge of the base flange and smoothly cut up through the centre of the raised flange. Steadily pull the rubber off the master (or the blade will damage it) as the cut proceeds up the side; ensure the cut is from front to back,

After much sweat and toil the rubber is removed, cleaned and secured in the case ready for casting.

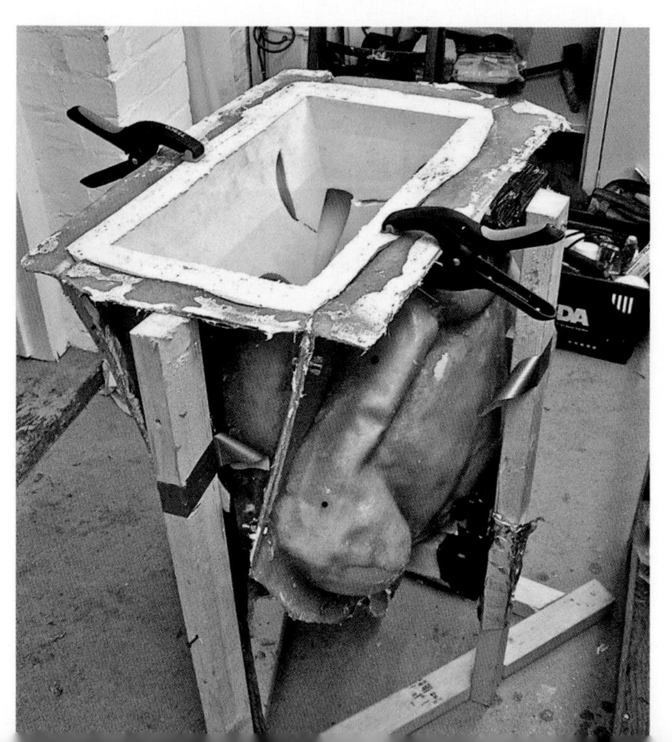

do not hack it up as it needs to sit back together comfortably. Lift off.

Casting

Clean all loose material off the case sections, trim any whiskers (where the rubber has seeped into gaps on the case or come through air holes), so that the rubber is neat and functional.

Loosely assemble the case and base with the bolts, slide the rubber inside and position it exactly making sure the cut sits true (for some complicated cuts it is okay to smear a mastic on to temporarily stick them together while casting; do not use silicone as it will bond with the rubber and eventually cause distortion), then do the bolts up tight, ensuring that at the end the rubber is a perfect fit.

Mount the mould in a tub so that it can be filled without falling or moving, and stablize it by pouring sand around it; more experience will result in the design of the casing incorporating its own base if possible.

Wet or Dry

This replica will be made from white cement (WOPC), sand and stonedust to match a buff limestone; WOPC is the best binder for this unless using the natural cements. A wet cast is exactly that; the concrete (or for sales: composite stone!) is made into a sloppy mix, poured in and cured. This results in a dense impervious finish to the replica that really does not resemble stone, more a clay finish as the fines all migrate to the surface for a smooth impermeable finish. This can be worked on afterwards, but personally I have never liked this casting method, but many people do; so here are a few tips to improve the standard.

With a nylon brush, reach into the mix in the mould at stages in the pour, and brush the surface of the rubber; this will stop air bubbles being trapped at the surface and causing blemishes which give away the fact it is a casting. To reduce the weight in the mix use lightweight expanded clay aggregate (or LECA – hollow fired-clay spheres of varied sizes).

Is This the Real Thing?

The problem with concrete replicas is that they do not look, feel or act like real stone. The following is a method based on experimentation, experience and mistakes that can give a pretty good result and has even fooled people into believing the casts were real stone; it looks right, the texture feels right and, being porous, will wet and dry similarly to natural fine-grained stone.

Procure a bone-dry fine sharp sand that, when viewed in bulk, is near to the desired colour, some dry stonedust and fresh WOPC. Use a strong mix here, about 2:7:1 (WOPC: sand: stonedust).

Colour

Mix these dry and look at the hue and speckle. This is what the finished colour will be, so, if desired, alter aggregate ratios and even add materials to improve the look; this cannot be given exactly here.

Wet cement castings do not weather like real stone. The shrinkage of this material also creates fractures that draw in moisture that causes the ferrous rods to rust, expand and destroy the piece.

A pair of finial tops wet-cast to match Coadestone, to replace missing sections on two gate finials.

GOOD VIBRATIONS

For the strongest castings (in concrete, natural cement or gypsum plaster) it is best to get rid of all the air and cause the mix to settle into every nook and cranny; just pouring in and a bit of tamping will not do this. Vibration is needed to move the air up, remove blems (bubbles that cause blemishes) from the surface and encourage liquids through. Pokers can be hired to do this and are advisable for bigger casts; alternatively, for smaller casts ingenuity can help.

HOME-MADE AIDS

Get an old jigsaw blade and tack-weld to it a piece of stiff wire or thin bar long enough to get to the bottom of the cast. Mounted in the jigsaw this can be worked up and down in the poured mix; this is a bit messy with the splashing of the mix. If an air hammer is available, hold against the sides of the case on full power to set up the vibration through the mix.

A vibrating table is simple to make and is great for small work and plaster casting. Mount strips of batten, for rails to stop moulds sliding off, around a decent piece of board (18mm ply or similar) of about 60cm square, but use what size is available. Solidly fix sturdy legs at each corner and screw half a rubber ball or doorstop to the bottom of each leg. On the underside, fix really securely an old electric motor that has a shaft protruding (water-pump or similar), attach a weight to the side of the shaft – perhaps a folded strip of lead and jubilee clips, but variations are obvious. When rotating this will vibrate the whole table (it will move about the floor, so be on guard) and thus the mould set on top of it. This is highly 'Heath Robinson' in approach so adjustment and improvement will be necessary as the contraption reveals its quirks and problems. You may come up with a better method yourself.

Reinforcing is needed for certain types of casts, especially in this case where it will be placed high above a street; if the worst happened and it broke, the pieces would not become detached.

Texture

Pour and work the dry mix through a sieve of about 1–1.5mm aperture. The finer it is the better it will pack, though for variation of texture this must be considered flexible; trial and results will help here.

The Casting Process

Here we drift into shed science, as the variables in materials and desired results preclude any absolute rules, so if changes or mistakes are made, or the result is not as expected, work out what may have caused it and adapt accordingly. For a standard bucket of dry mix, have a container of water and dribble in about a third of a litre, mixing thoroughly. Take a handful of the mix and squeeze it; if it retains the shape of the squeeze it has enough liquid in. Dribble and mix to get this right.

Packing It In

Make a couple of smooth wooden tampers that will fit down to the bottom of the mould with as large a footprint as possible to compact the mix, and place the mould solidly on the ground so that it cannot rock. Scoop mix into the mould and fill for about 10cm or so. With the tamper push the grains hard into all the corners and start compacting by beating the tamper with a mallet; with practice this will get a solid layer in. Be aware of the tendency for the rubber to bounce and break up the compacted section, so work to the limit of this. There will now be a smooth-topped layer in the bottom of the mould. With whatever effective tool comes to hand (for example, a batten with a saw blade in the end), score the surface to key for next layer. Repeat this process up through the mould, ensuring each layer is compacted to the same extent. At the top cut a board to fit snugly in the top and do the last compaction with this. For some castings, such as this example, where narrow portions could make it fragile if knocked, insert some reinforcement (here use 10mm stainless steel rebar), cut overlength to provide a handy fixing dowel at one end, pushing it firmly and smoothly down through the centre of the cast; give a little bit more compaction around this if needed.

The Cure

This needs to be left alone while a semi-set occurs; this is when it is solid to the touch (but do not dig into it) – it tends to be about two to four hours.

Once it seems to be set, gently pour beakers of water into the cast. Do not flood it or wash away the mix, but continue until the cast is just about to overflow, indicating that all the pores are filled with water, and so the crystallization of the mix can take place – once again the time depends on many factors, so leave for at least two days.

The Result

Demould as before but without the need to lever flanges. If all is well there will be a perfect cast. If it crumbles then it needs to cure longer. If projecting edges are granular or crumbly then better packing is needed; these are not really problems, just learning steps.

The finished article will have a good even colour. The texture will be stony and, when water is poured on, it will soak it up, showing good porosity; this will fit in well with similar stone and due to the texture and porosity will weather up and attract the same soiling and growth as natural stone.

The only downside, and this is purely from a conservation ethics stance, is that it is a lot tougher than stone and will not erode as easily; this is a cost that can be borne.

Variations

Use different stonedusts for other stones. Crushed Portland dust from 1mm down as the aggregate is very effective. For a sugary-surfaced marble effect (which matches exterior weathering of marble as it loses its polish), get hold of the marble dust used in air abrasive cleaning. Variations in Yorkstone can be suggested with patches of browner mix thrown into a layer – do this at an angle to give a natural bedding effect.

Redder stones are harder to match, due to the lightening effect of the WOPC, but try crushed brick dust (of the reddy-purple colour). Experiment whenever the opportunity occurs, using leftover materials and try to keep a notebook of results. Enjoy yourself.

Brush-On Moulds

With some masters that have a (preferably flat) back nearly the size of the face (a shield or attached baluster, for example) that can be laid flat on this back without deep undercuts or ledges that are hard to access, then a brush-on mould will cut out all the clay work and can be more economical in rubber usage. To start, the master here will be a rectangular block with an attached (half) baluster, so the case will be three sections with joins along the centre and at one end – it will make sense further on.

Dry-cast balusters kept under polythene and stood in water to cure properly. The texture and feel of these is very close to stone.

Set Up

On the ubiquitous smooth board with plenty of margin, set the master on a very thin bead of non-setting mastic around the perimeter of the back to stop ingress of silicone beneath the piece; if it is uneven on the back, try a strip of double-sided draught excluder, with an overlap at the ends; press the master down firmly to take up the contours. For backs with large undulations or similar, cover the back and up the sides a bit with foil, then pour a stiff plaster pad onto the board and press the master onto it, squeezing to get it as close to the baseboard as possible. When the plaster is almost setting, scrape the excess from around the master, take off the board and smooth the edge. Let it set.

Roll out a long bead of clay and build a 5–10mm fence completely around the master, 50–60mm away, to stop rubber pouring off.

Laying It on Thick

While this part is about brushing on rubber, the most effective routine involves mixing up a small amount of rubber (guess-timate out how much) and pouring it over the master, to get a continuous contact with the surface. The rubber will flow off the master and onto the board, hence the fence; repeat a couple of times over an hour or so and eventually a skin will form. Let this get into a good cure.

Now mix up enough rubber to completely cover the piece with a thickness that will not tear when pulled off. Add the right amount of thixatropic additive and mix until it thickens, then apply liberally, to encase the whole piece. The poured rubber will have formed the flange, so smooth this onto it; the rubber will not smooth well as it is so sticky – it will end up resembling Christmas cake icing. To get it smooth decant some silicone fluid into a beaker and, using a brush, level the surface of the rubber, remembering that having to exit the case without snagging will determine the shaping. Trim the feather edging off the flange when it has cured.

The Case

A GRP case in one piece or in sections can be made as before, so that is a known; this one, for an alternative, will be a sectioned plaster case.

Decide where the joins in the case will be, knowing that ease in dismantling and solid strong corners will improve its life and usability; draw these with a marker on the rubber as a reminder. The sections will need a way to be held together firmly when turned over for a cast and possibly vibrated; a flat top (handy when it becomes the base) is also part of the design criteria.

Large brush-on mould under construction; note the lugs adhered to the silicone for locating in the case.

Fencing

The plaster will be poured in and will not be thin and contoured, rather three blocks that together make a blocky sarcophagus, the mould safely reposing within, so another version of case making is needed. For the first case section construct a hardboard (shiny side in) fence around one end of the master following the flange, about 75–100mm out and 50–75mm taller than the rubber. Fix blocks to the baseboard around the outside to keep it in shape and position.

Now make a thick clay wall for the joint side across the rubber to the fence; buttress this, smooth the inside and in a couple of places press a concave depression (marble, spoon, thumb) for locating lugs. Do not forget to run a clay seam along the base of the fence to stop leaks. Look at it critically now, as it will have a large volume of plaster poured in, creating pressure and flouting weak points; work out how it will perform and modify as necessary. When satisfied all is ready, give a spray over the interior with release wax, WD40 or soapy water; all are good and there may be others hanging around on the shelves.

Smoothing the surface of the brushed-on silicone.

Plaster Mixing

Mixing plaster is very simple, with little to go wrong. The only things to worry about are the setting times and hardness of the plaster. Plaster hardness is categorized from A (softest) to D (hardest), and when ordering note the setting time, which tends to be shorter for the harder plasters. For cases a softer plaster will do, so buy general casting plaster (large sacks, not the hobbyist size!) with a decent working time, and a couple of sealed tubs to keep it in; it will degrade through contact with moisture in the air.

There will be a specific water-to-plaster mixing ratio noted somewhere, so go with this as a starting point, especially for the harder plasters; but here we will cover the rule-of-thumb workshop method for mixing casting plaster. Guesstimate the amount of plaster needed and put this amount of clean water in a mixing tub, then scoop up plaster and sprinkle it into the middle of the water; continue doing this until an island pokes above the surface, and remains there. Art students among you would now dip their hand in and start mixing. It is good fun but messy so just use a paddle, stick or trowel until it is the consistency of single cream; pour this carefully into the space and let it set.

Remove the clay wall and the fence, and put another fence further down to encompass the next section; for some it will be possible to build a completely enclosing fence in one go and keep it in place until the case is complete. Build a clay wall from the plaster block along one side of the centre line to the fence, buttress it and put in the dimples/lugholes, seal the base of the fence and apply the release agent making sure the plaster block has a good covering. Mix and pour the plaster as before, aiming to get it on the same level as the previous block, but do not worry too much as it can be scraped down afterwards; let this set. For the last section, remove the clay wall, as it is not needed now, and carry out the plaster pouring as before.

Using the Mould

Take down the fence and, with a drag or similar, clean off the top to a level flat surface. Take the sections apart by levering the joints and clean off any extraneous bits to the sections. Pull the rubber off the master, and place it in the reassembled case; to keep the case together put a ratchet strap around the perimeter and tighten. Make the cast as before.

Plaster Modifications

For big cases or plaster casts, a lot of plaster needs to be mixed and the chances of it setting during this are high, so add a retarder (tri-sodium citrate) to extend the working time.

Where larger, thinner or more delicate plaster casts or cases are required, move the plaster performance towards that of GRP by reinforcing with scrim or chopped strand fibreglass. To strengthen and harden the plaster there are polymer additives available, though a good dash of PVA glue in the water will toughen it up.

Brushed-on mould and GRP case, showing the clean surface, free of snags, which aids demoulding.

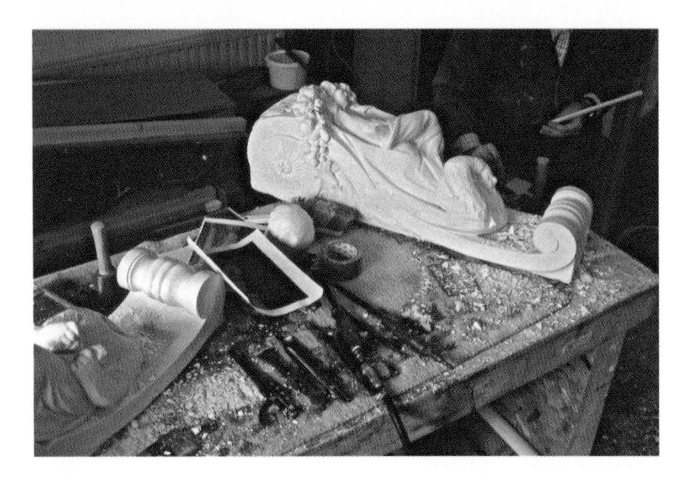

A massive marble fireplace had been stolen; as it was one of a pair we had a model to cast up a new one as the expense of carving was too great. It had allegorical figures of the Four Seasons, so casts of the existing two were remodelled to replicate the missing ones. Here, heads are being worked on.

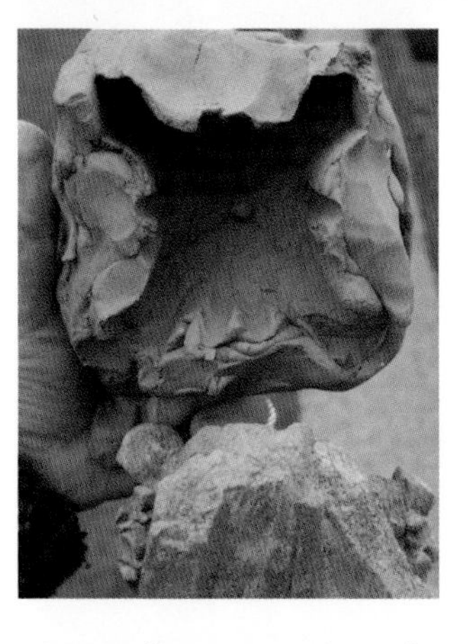

Here, a clay squeeze is made to get the shape of a broken finial so the replacement parts will match in to the original.

Part of the same project, the small leaf is cast from a squeeze to get the style of modelling. A mould of the new-modelled clay element of the missing piece was made and repeating leaves were cast, then built into replacement masters that were cast as one piece.

Worked up and ready for more moulds to be taken.

One of the waste-casts of the finished pieces, also a plaster cast of a Portland corbel showing good undercut and detail. The whole surround was cast in a modified plaster designed to take a polish and resemble white marble.

Squeezes filled with plaster on site.

As large flat or long panels of plaster dry out they may distort, which is not a good thing; adding a trowel of hydrated lime to the water (this is a result of shed experimentation, so no figures yet!) prevents this.

To add colour to plaster, use pigment mixed in (hot) water. The plaster will be much paler than the liquid when finished so add plenty of pigment to the water before the powder goes in; keep this in suspension or the pigment will fall out.

Jesmonite

This is a modified plaster in an acrylic resin (not dissimilar to what was proposed above, so a viable alternative to GRP or plaster) used in production of casts for art, architecture and film. It is obviously more expensive, but due to its strength can be used to make lightweight pieces; it is worth getting hold of some, trying it out and comparing it to homemade modified plaster.

Taking an *in situ* mould of a bracket; the stone has been covered in glazing silicone (easily bought in tubes from hardware shops) and protection put up for making the case.

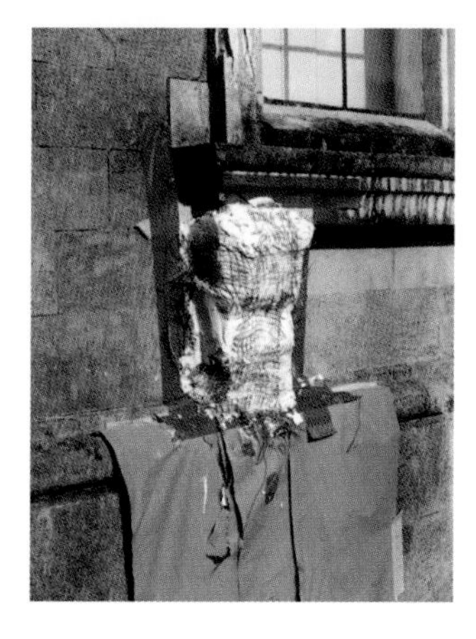

Building up the plaster case and reinforcing it with hessian.

Lime casting of missing volute was made in a squeeze mould from a volute in good condition.

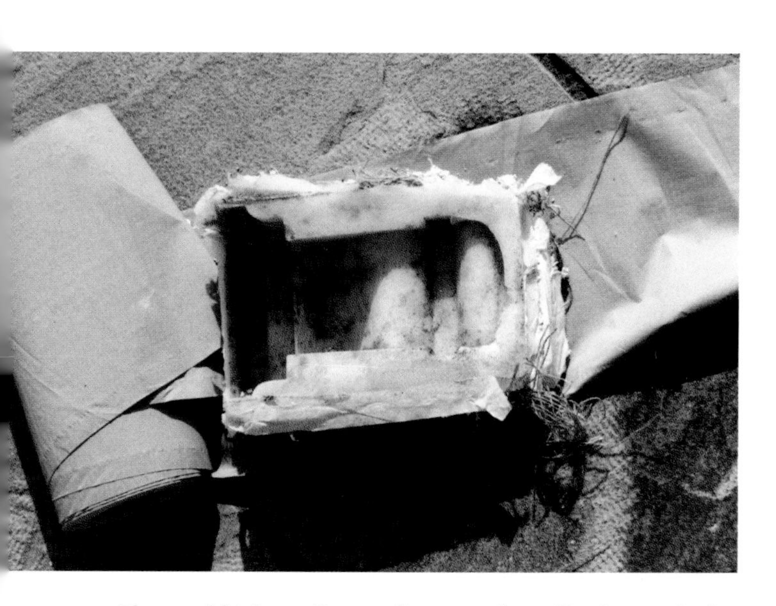

The mould taken off – a rudimentary but effective method.

In Situ Moulds

Occasionally the bit to be replicated is still attached to the building. Resist knocking it off; instead learn how to apply a squeeze mould.

A Simple Approach

The most basic way of getting a copy of a small detail is to literally squeeze rubber onto it, peel it off and cast in; practise this at home to become aware of limitations and problems. There are fast-curing silicones available for this either as in tubes or as putty, or by using a specific catalyst the cure time of (thixatropic) moulding silicone can be drastically reduced. Another method, that requires access to the piece the next day, due to cure time, is to use silicone air-drying mastic in a tube. Squirt this onto the piece and press in with a palette knife or have a bowl of soapy water and dip a gloved hand in then hand-press it in. Build up enough thickness, so the mould retains its shape without support when pulled off. If the piece is bigger then a case may be needed; this can be plaster or GRP depending on personal preference and experience. For a one-piece case, build up the rubber, sloping it so there are no catches for the case. Mask off the surrounding area to prevent staining from spills or drips – polythene sheet and gaffer tape are the basics for this.

The Case

GRP is already covered, though it may be trickier on site and warm weather could speed up the cure, so be aware.

Plaster (of Paris) is cheap and relatively easy to use. With a prepared piece covered in rubber, mix up enough plaster to use before it sets (experience will tell here) and apply to the piece as it stiffens up. Reinforce especially around the rim with scrim, plaster bandage (this can be used on its own if possible), chopped strand or a piece of fibre mat (spray this with a bit of water first or it will not accept the plaster), laid on the damp plaster then covered with another smear of plaster. The properties of Jesmonite make it a good proposition here,

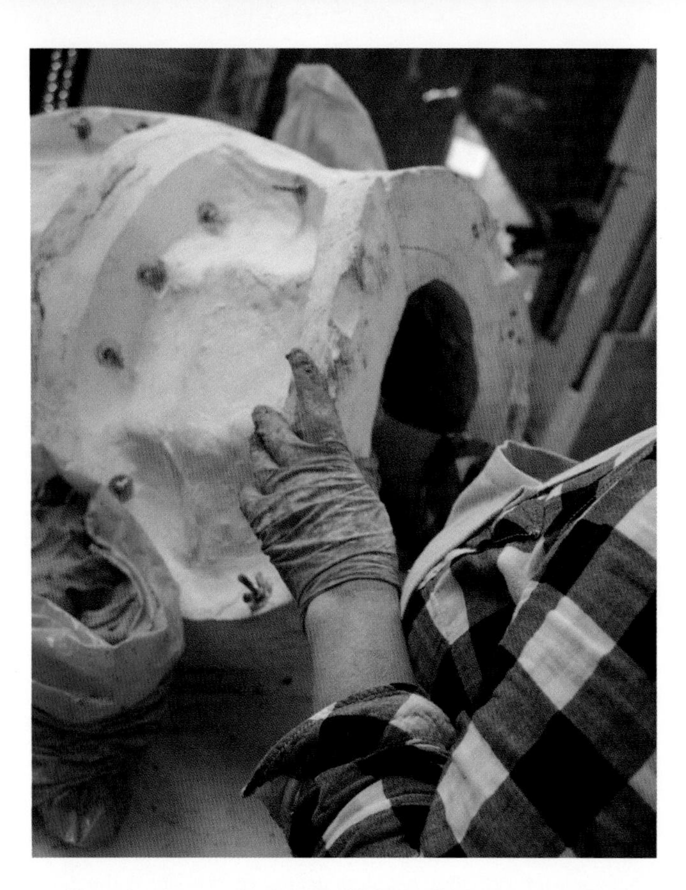

Packing clay into a plaster mould for making new Coadestone pieces; the detail and undercut is added when the clay has hardened and shrunk a bit.

bridging the divide as it does between GRP and plaster. When you are happy it is set or cured, the case can be levered off; be careful with the plaster, as it is brittle stuff. Peel off the rubber and start casting. Multi-section cases can be built up *in situ*, similar to the workshop method but there will be less ease of access, so always have a workable design before making and remember to preserve the original.

Missing Part Production

Occasionally there may not be an original to copy, perhaps where a chunk has been knocked off a carving leaving a ragged edge and the requirement is to keep all the original material (so no cutting back for a mortar or Dutchman repair) and get the piece looking whole again. The simplest way is to model up the missing bit in clay and, before it shrinks, make a mould to cast anew. The advantage here is that the joint can be almost perfect; this is almost a museum technique but has been used to good advantage on small carved pieces on monuments or sculpture.

Clay Work

Set the piece to be worked on so that it is steady and drill one or two small holes in the joint. If it is for a hanging piece where gravity may be a nuisance, set long dowels in; when the bit is to be supported by the main body, small pegs just to lodge it will work – this is a site decision, so make it. Powder up the surface of the joint with some fine stonedust to stop the clay sticking and distorting when removing it; get some seasoned clay (stiff and good to work, not straight from the bag), build onto the joint and work up the missing detail *in situ*.

The One-Off Mould

When happy with the quality of the mock-up, let it harden slightly but not shrink, and remove it without changing its shape. Depending on the size, have a tub that the clay can fit in comfortably and put in enough (soft casting) plaster to come up halfway on the piece (this will be the back part of the mould). Put a 10mm-thick plug on the modelled piece and place it in the plaster face up so it has plaster behind and around the edges. Gently tap the container to get the plaster level and into the nooks, then, as it hardens, scoop out a couple of locating dips with a spoon or similar and let it set properly. Apply a release coat to the plaster and clay, pour another layer of plaster on to completely cover (allowing at least 10mm thickness at the highest spot), and let it set.

Last Stage

Remove from the tub and split the case. If it is a complicated shape the clay will break up; this is okay so do not panic. Peel out all the clay, give it a wash if bits have stuck inside and apply a release coat to the inside. The plug on the back will give a pour hole for the casting mix; if a larger one is needed work out where it will go without spoiling the cast when the stub is cleaned off, and cut in with appropriate tools. Fasten the case together tightly with a strap, cable ties or clamps and pour in and work the cast mix as normal. When it has set completely, break off the plaster by cutting and careful chiselling to release the cast; use picks to get it out of the nooks, being aware of the surface of the cast at all times. When out, carry out any necessary work, such as creating texture and drilling dowel holes, to the cast and fix to the original.

A replica wooden angel, made from plaster and given a finish based on pigments and wax; it was not painted, as budget was very tight for this project.

Clay packed onto face to take a squeeze mould for one use only.

The plaster cast of the face was used to work the Dutchmen exactly to fit and then be carved in; unable to make it any lovelier!

Finished and cleaned.

Vandalized face of William of Orange statue in Brixham.

Here, in less restricted days, carving the face scars to have flat planes to allow small Dutchmen to be fitted. Note the sensible shoulder bag for tools, etc., and the protection under the ladder.

SCAGLIOLA

Everybody wants to live comfortably, perhaps in a palace lined with marble columns and sumptuous fittings. Obviously this cannot be for all, as marble is scarce, expensive and hard to work. Luckily, many centuries ago, enterprising craftsmen in Europe developed a material that could bring the touch of marble luxury that was needed to enrich churches and grand buildings without breaking the bank: this is scagliola.

Derived from the Italian term *scaglia*, meaning 'chips', scagliola is a mixture of plaster, animal glue and pigments, made in such a way that when finished it can resemble marble or semi-precious stone. It has been used to make most architectural elements, such as columns, brackets and sculpture, usually in a mould traditionally made of plaster to stand up to the pressure of packing it in, for the solid cast method known as Bavarian scagliola. It can also be made into blocks, or turned for finials or urns, and can be a good imitation of *pietra dura* (literally, 'hard stone') by inlaying it in other colours or styles, into cutout designs in plaques or similar and then polishing back.

This section is included in this book to introduce this lovely material that is great fun to produce and that has appealing results. The connection with repairs is that it does break down and not many people can make it (the apprenticeship is about seven years!). It is occasionally used to repair marble which, although it takes a bit of practice and confidence, is good experience.

THE RECIPE

Here I present the mix written down by myself (with alternatives if necessary) while attending college in Venice, so it is translated from my pidgin Italian but fortunately lacks the pasta and wine stains.

Materials

- Good-quality gypsum or alabaster plaster of Paris (I tend to use a medium [A–B] white plaster)
- Glue water – dissolve 450gm pearl or animal glue in 10 litres of water overnight
- Various powder pigments (oxides or earth) in colours needed
- You will also need plaster retarder to give time for mixing and working – try out several!

Tools

These all need to be scrupulously clean and polished smooth.

- Long chop knife or a plasterer's mitre or steel straight edge with sharp grind (this can also double as the leveller)
- Cutter, which is a square-ended trowel or a big paint scraper
- Small, good-quality trowel with stiff blade and rounded end
- 12mm paintbrush, bowls and paper cups, disposable gloves, sponge and bucket of clean water
- If possible, a helper
- Polishing kit – wet and dry paper or diamond pads

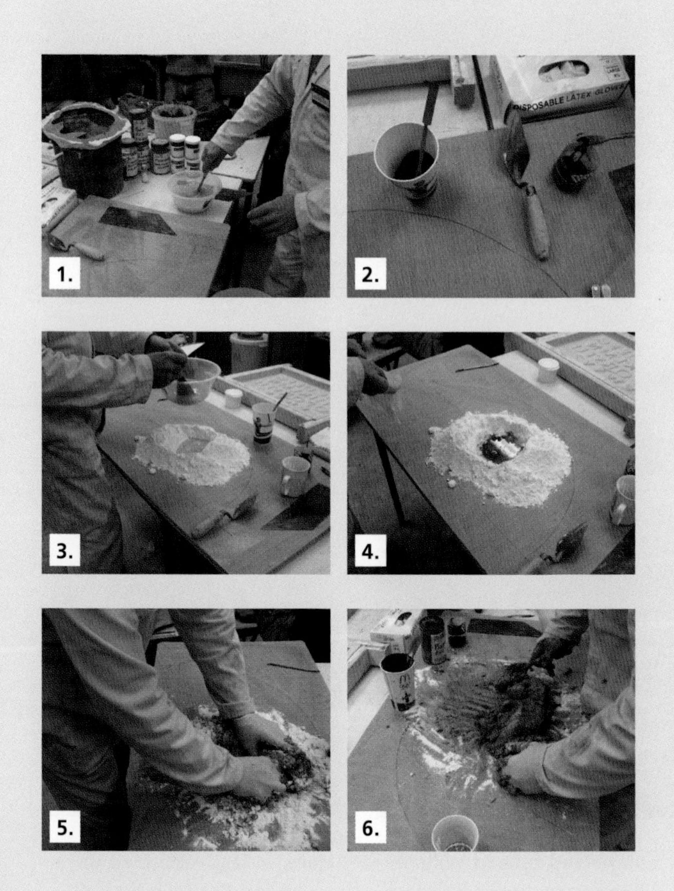

Get a mould to work with or, for experimenting, try old flat-bottomed containers that the solid cast can be removed from, providing you do not mind breaking them. Or make a flat tray, screwed together with sealed surfaces, to make just a flat slab version.

Method

Set out all the ingredients and tools on a large smooth surface such as a section of kitchen worktop, giving plenty of room to work; the proper workshops have polished granite tables, so if you can get something to match this, do. The basic methods will be described and accompanying photos show the stages, but there are no hard rules; the best way is to practise and experiment with colours, marbling and veins.

Have a pile of damp sheets of kitchen roll for wiping hands and surfaces.

- Add the retarder to the glue water.
- Mix pigments with hot water in plastic cups. This ensures they mix well, though variations and accidents can be interesting, so do not get too finicky.
- Pile a ring of plaster onto the table and pour glue water into the middle and add a measured amount of pigment juice. Keep a track of quantities and colours in case it needs to be repeated; also you start getting an idea of what is needed in terms of strength and combinations to match colours.
- Start mixing and kneading until the dough is completely coloured and the consistency of shortcrust pastry.
- Cut into four pieces and add a quarter-measure of pigment juice to one, half to another and three-quarters to the third to get the nuances of colours. Obviously this can be altered or increased for other variations – experiment!
- Make rolls of these.
- Mix about a fifth of the first mix of runnier clean mix with the vein colour; use your helper or put on clean gloves – this will get messy.
- Dip your hand or the brush into the vein mix then smear it unevenly between the rolls and form them into one large roll.
- Slice into sections of 20mm or so and stack them up.
- Place these cut side down into the mould in a pattern that replicates the figuring in your chosen marble so that there are no gaps and press it in really hard with the trowel; the aim is to push all the mix to the face of the mould or there will be holes which need filling afterwards – although for an eroded appearance this can be actively tried for.
- Leave to set for at least a day, or more if the retarder slows it down. When it has stiffened up a bit it is worth trimming all the edges where it was squeezed out of the mould, or to square off the back – it will be too hard to do this afterwards.

Variations

- Flatten the roll or make a 'layer cake' instead to get long markings.

- Layer various shades of green like uneven puff pastry and roll out to make long thin strip sections when cut; lay so the stripes undulate across the mould for a version of malachite.
- Make lots of thin rolls first to get a speckled appearance.
- Lay the sections of roll in uneven narrow strips/groups of various-sized tube sections angled across the mould and use a plain colour mix to fill in the large background areas.
- Use a couple of vein colours.
- Try to copy a real marble like Verona red and keep tabs on how it works.
- American scagliola is made with a liquid mix and has silk or cloth soaked in pigment drawn through it to make veining; this should be self-explanatory – once again it needs to be tried rather than talked about.

Use your imagination and think of applications; the photos of the ammonite and lettering are Portland stone with the design cut out, scagliola set in and then polished.

POLISHING

- Demould when it is set really hard.
- With plenty of fresh water start polishing it with 600-grit wet and dry paper to get an even surface finish.
- Rinse and let it dry, then check for holes that need filling.
- Mix up small batches of appropriate colour mixes with putty consistency, brush a little diluted glue water into the holes but do not let it stay there.
- Pug out all the holes and missing bits, filling them proud, and let these set.
- When ready repeat the 600-grit polish, then rinse and throw away the polishing water (any bits will scratch the surface and ruin it) and get fresh.
- Repeat the above with grit grades 800, 1000, 1200 and, if wanted, finer, always washing off and using clean water.
- Let it dry, then wax and polish; it should now feel and look like marble – perhaps the first steps to a new hobby.

For flat areas use blocks to keep it level and make shaped blocks for mouldings.

Diamond pads never wear out and can be had in grades up to a fine buffing finish so they may be used instead, but try not to rub an undulation in as the beauty of marble is its quality of surface level and finish.

APPENDIX 1: PUTTING IN A SLAB

Slab

A well-worn and mellow stone floor is often the first line of defence against damp in the ground; many historic examples were laid directly onto earth and can break up due to the wear and tear of surface traffic. Pulling out and replacing single stones is a straightforward task; it just needs the usual careful approach. Here follows a related example – the setting of a commemorative plaque, beautifully designed and cut by lettercarver Andrew Whittle, in the porch of Sherborne Abbey. With such an important piece, quality of work is paramount, but really no more than any other task for the artisan.

Mark out the dimensions and cut out the joint with a diamond blade disc cutter.

Lay straps into the hole, and place the new slab in to size up.

Using protective padding, carefully get a lever under the edge and lift up; as the edge is raised, slide wedges underneath to free it.

Lay a dry gritty mix into the hole in a ring around the edge and across the middle, banked higher to allow compression which will push the bedding down into the spaces.

Mask up the new slab, wet down the stone and trickle a very dry mix into the joints, packing down as you go. Leave it to stiffen up and brush back to level.

If working on a floor that is also a ceiling, always consider the chance of collapse and take precautions. Acrow props, with soft padding to prevent damage to the stone, were placed to avoid any collapse as the frame above was removed.

Vault

The delicate medieval vault in St Mary's, Brixham, was prone to spreading and had been the subject of Victorian intervention to ensure the keystone would be held in place, by attaching a metal pin to a wrought iron frame situated above it, should things get worse. Unfortunately the pin and the metal cramps across the backs of the stone were rusting away due to the ingress of water and causing their own brand of damage. My design for a replacement support entailed fabricating it *in situ*; getting the metal pieces up was relatively easy compared to lugging my massively heavy old-fashioned welder up a ladder into the enclosed space!

Lining up the new sub-frame prior to fixing and welding. The extent of rusting cramps is shown by all the shaped holes in the stone. The stones were now fixed to the new frame by studding set in their backs.

The medieval vault had had an iron support put in to hold up the boss-stone. This support was resting on pads on the corners of the vault.

The whole ensemble was now in place. As this involved welding it was necessary to get a hot-works certificate. Be aware that any sparks or hot metal can fall into a nook and ignite debris or cause dry timber to smoulder away unseen.

APPENDIX 2: KEEPING IT TOGETHER

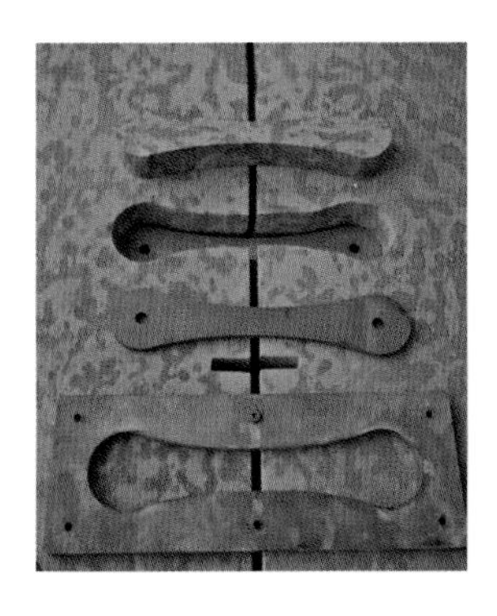

The set-up – the template, with fixing holes, the bone-shaped 'cramp' and the cover block.

Following the template with the grinding bit to remove the centre and level out the bottom of the cut.

Rather than relying on the old style of dog cramps that were traditionally used to hold stone in place, here is a bespoke system, made using modern technology, that provides greater restraint and is easier to accomplish.

All the components are designed using CAD and cut out using waterjet so that the template is just bigger than the cramp and the stone cover.

The metal link shape (cramp) is all curves (and can have a pin dropped through the centre for extra keying if desired) which, when set in the correct shaped recess, spreads out the pressure evenly along the outer edge of the cramp. It is stronger as it is set deeper than standard cramps (some rusty ones are shown earlier in the book), so there is more stone to prevent a shallow spall. The stone cover allows you to blend in the work and is more durable than the usual thin layer of mortar that tries to stick to the metal of traditional cramps.

Using a snug fitting coring bit to take out the ends.

Cutting out is simplicity itself. The template is mounted over the two stones and fixed in place by screwing into Rawlpugs set in the joint. The size of the cut-out should be such that the ends can be exactly bored out with a diamond core bit mounted in a small angle grinder. The middle is then removed with a diamond routing bit with a guide bearing on the shank to follow the template shape. The recess has a nice runny grout put in so that when the steel and stone are pressed in tightly, all the voids are filled.

The whole process takes about fifteen minutes, is always more accurate, consistent better in quality and at less cost compared to the old style.

Cover block dropped in to show the depth of the cut, done well and with a cured mortar this should blend in and be very durable.

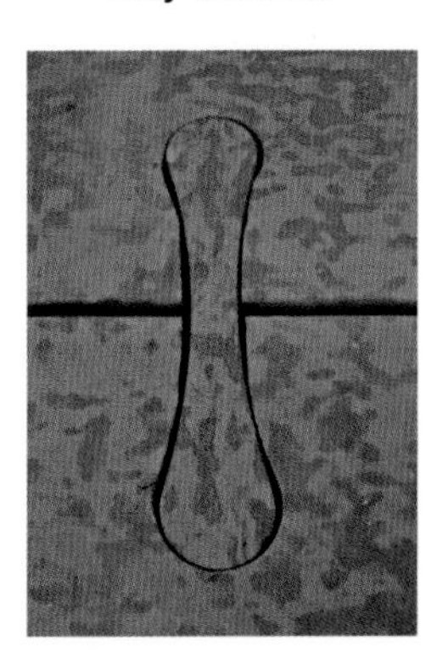

GLOSSARY

Aggregate: Particulate material used in construction.

Armature: Structure built to support from within.

Arris: Projecting edge where two surfaces meet.

Ashlar: Masonry construction of blocks with worked, even beds and joints (generally under 6mm) and set in horizontal courses. Stones within each course are of the same height.

Banker: Stand or bench for working stone on.

Bed: Horizontal surface of a stone; position where fixed.

Bevel: Sloping surface.

Biocolonization: Situation where living organisms inhabit or colonize.

Bonding: Overlapping of stones so that vertical joints are staggered.

Boss-stone: Top, central stone of a vault.

Bouchard: Hammer with grid of pyramidal points on either end of the head – similar to a steak hammer.

Box trammel: Instrument to scribe in a line parallel to an edge.

Calcine: To reduce, oxidize or desiccate by roasting or applying great heat.

Calcite: Mineral made of calcium, carbon and oxygen ($CaCO_3$); the principal carbonate component of limestone, chalk and marble.

Cast: Object made in a mould.

Cation: Positive ion.

Cementation: Diagenetic process by which the grains of a rock are bound together by minerals precipitated from associated pore fluids (for example, quartz and calcite).

Centring: Frame to support arch during construction.

Chalk: Very fine-grained white limestone composed principally of microscopic skeletal remnants known as coccoliths.

Chamfer: Symmetrical sloping surface at an edge or corner.

Chert: Granular microcrystalline to cryptocrystalline variety of quartz.

Chevron: V-shaped mould.

Cimente-fondue: Fine concrete used for sculpture and decorative work.

Claw: Serrated edged chisel.

Colloidal: Describing a substance microscopically dispersed throughout another substance.

Core: Material within a wall, usually rubble and mortar, filled in as the outer skins are built.

Cramp: Method of tying two stones together across the joint.

Cross-section: View through the side of an object.

Crypto-florescence: Hydrated crystal growth beneath the surface.

Crystallization: The growth of a crystal as it absorbs moisture.

Cusp: Projecting point of curve(s) in tracery.

Details: Dressed stone components of a building that are not walling.

Diagenetic: Process of change.

Dogleg: Continuous line that has one bend.

Dowel: Length of rod, usually metal, used in a hole for location or restraint.

Draft: Narrow worked area along a stone to set a height.

Dressed: Stone that has been worked.

Dressings: Worked stone such as frames, quoins and the openings of the structure.

Dunting: Lines in stoneware or terracotta that look like vents, a result of the firing process.

Efflorescence: Hydrated crystals on the surface.

Entasis: Concave curve added to a shaft to prevent it looking too waisted.

Escutcheon: Shield-shaped carving or design.

Extrados: Outside line of an arch.

Face-mould: Templet that shows the shape of the stone from the front.

Ferramenta: ironmongery.

Ferrous: Of iron or mild steel, susceptible to rusting.

Ferruginous: Containing iron minerals usually in the form of an iron oxide, which can stain.

Fillet: Mould (small) with square section.

Finial: Carved or ornate ornament atop a building or plinth to terminate vertical work.

Fixing: Building stones into a structure.
Flange: Extension used to rest or secure sections together.
Flint (or chert): Hard, resistant beds or nodules composed of cryptocrystalline silica; flint is restricted to nodules and beds that occur in chalk (Upper Cretaceous) rocks.
Fluting: Moulding design to column or pilaster.
Foil: Opening between cusps on tracery.
Form: Board or structure against which mortar can be built up or stones supported while fixing.
Fossiliferous: Containing fossils.
Freestone: Stone that can be cut and shaped in any direction without splitting or failing.
Friable: Crumbling and dusty.
Gallet: Small stone or pebble inserted into joint or mortar.
Grotesque: Animalistic carving on building.
Grout: Liquid material used to fill voids within an object or masonry.
GRP: Glass reinforced plastic, commonly called fibreglass.
Guesstimate: Figure based on experience, arrived at without calculation.
Hawk: Small flat board with vertical handle underneath, used for holding mortar.
Homogeneous: Without any different parts; consistent.
Hood/Label: Protruding moulding that diverts water from windows and doors.
Hydraulicity: Ability (of mortar) to set under water or moist conditions.
Hydrophilic: Describing molecule or other molecular entity that is attracted to, and tends to be dissolved by, water.
Induction: Talk or presentation covering all aspects of a site.
Intrados: Inside line of an arch; soffit.
Jamb: Vertical sides to an opening.
Joggle: Indent or channel cut in joint section of stone.
Joint: Line where two stones are put together.
Keying: Roughened or scored surface to provide grip or purchase.
Keystone: Top stone of an arch, locking the structure into place.
Lath: Wooden strips fixed in rows to support plaster and render.
Leaching (out): Mineral solution exiting masonry.
Lesbian curve: Flexible mason's rule, originally made of lead.
Limestone: Sedimentary rock consisting mainly of calcium carbonate ($CaCO_3$) and containing grains such as ooids, shell and coral fragments and lime mud; often fossiliferous.
Lintel: Stone spanning an opening, supporting the masonry above.
Lithology: Description of a rock based on its mineralogical composition and grain size; for example, sandstone, limestone, mudstone, etc.
Maquette: Reference model for making a larger piece.
Marble: Metamorphic stone.
Mastic: Plastic (mouldable paste or liquid) material used to seal or set dowels in.
Matrix binder: Cement that holds the grains of a stone together.
Merlon: Capping and side stones to castellations.
Metamorphic: Describing rocks that have been subjected to heat and/or pressure, which have caused changes in their solid state.
Miscible: (Of liquids) forming a homogeneous mixture when added together.
Mitre: The meeting of two similar mouldings, either internal (going into a corner) or external (forming a projecting corner).
Modillion: Projecting bracket under the corona of a cornice.
Monial: Where a fillet follows a curve in tracery.
Monolithic: Made of one stone; term often used to indicate great size.
Mould: (Flexible) form in which setting material is placed to produce an object.
Mouldings: Dressed stone with a contour or section, either projecting or inset, to give emphasis, usually to horizontal and vertical lines.
Mudstone: Fine-grained sedimentary rock composed of clay and silt-sized particles.
Mullion: Vertical divider of a window.
Normal: (Of a striking action) at right angles to the face of the stone.
Ogee: Moulding with two (reversed) curves.
Ooid: Spheroid grain of calcium carbonate formed by precipitation (by algae) of calcium carbonate in concentric layers.
Organic (solvent/consolidant): Derived from oil or petro-chemicals.
OPC: Ordinary Portland cement.
Outcrop: Area where rock is exposed at the ground surface; the tors of Dartmoor, for example.
Overburden: Top layers of a quarry, consisting of loosely packed material.
Ovolo: Moulding of a quadrant (quarter circle).
Oxidize: Where an oxygen molecule is added to a material; typically rusting.
Packer: Flat piece of material to hold two objects apart.
Permeable: Allowing moisture to pass through.
Perp(endicular): Upright joint in masonry.
Petrographic: Pertaining to the mineral and structural make-up of stone.
***Pietra dura*:** Inlaying of stone into stone artefacts to create designs; originated in the Medici workshops in Florence.
Pitch: A strike in stone to cause a directed fracture.
Plaster: Wall covering of mortar.

Plastic: Describing material that falls between solid and liquid and that can be modelled using hand tools.
Plinth: Base courses or blocks of a wall or column.
Pluke: Small friable scar, usually in scurf.
Porosity: Ratio of the fraction of voids to the volume of rock in which they occur.
Poultice: Pulp material used to hold liquids in close contact with another material.
Pranked: Dotted randomly here and there.
Quoin: Corner (stone).
Radiating joints: Joints that go towards a common centre.
Rebar: Metal rod with protruding nubs.
Reduce: To stop the oxidizing process by chemically adding an oxygen molecule.
Reglet: Narrow band separating mouldings, panels or courses.
Relieving cut: Furrow to stop the force from another cut in stone travelling too far.
Render: External mortar covering of a wall.
Resin: Liquid material that when mixed with catalyst hardens to a strong plastic.
Return: Direction change of a moulding or wall.
Roll: Curved moulding, usually circular in section.
Rubble: Rough, undressed or roughly dressed building stones typically laid uncoursed (random rubble) or brought to courses at intervals. In squared rubble, the stones are dressed roughly square, and typically laid in courses (coursed squared rubble).
Rusticated: Describing where the joint is recessed back from the face, which may be worked in any style.
Sandstone: Sedimentary rock composed of sand-sized grains (that is, generally visible to the eye, but less than 2mm in size) bound together by silica.
Saturation point: Point where a porous object cannot absorb any more liquid.
Saw-lash: Ridged marks resulting from saw vibration while cutting stone.
Scagliola: Plastic material that sets hard and can be polished to imitate polishable stones.
Scribe: To score a line in stone with a pointed tool (scriber).
Scurf: Build-up of calcium sulphate as a product from acid rain attack on limestone, resulting in a brown to black layer of sooty particulates.
Section: The shape viewed when cutting through a stone or drawing a moulding.
Sepiolite: Naturally occurring complex magnesium silicate forming a fine clay with very high absorbency.
Sinter: To coalesce from powder into a solid by heating.
Socle: Waisted element supporting a finial, urn, column or bust.
Soffit: Underside of an arch, lintel, ceiling or overhang.
Softening: Material softer than stone used to prevent damage.
Solulize: Diluting or softening a material with a liquid.
Sorptivity: Measure of the capacity of a medium to absorb or desorb liquid by capillarity.
Spall: Shard of stone that breaks off.
Span: Distance of an opening from side to side.
Spatula: Small double-ended trowel-like tool used for pointing or modelling.
Springing line: Horizontal level at which an arch starts.
Strike: The hitting of a chisel with a hammer or mallet.
String course: Moulded course of stone that runs horizontally.
Stucco: Term for render, also for modelled plasterwork.
Studding: Metal bar threaded along its length.
Surfactant: Material that breaks down surface tension of liquids.
Tangent: Straight line that touches a curve without cutting it.
Templet (Template): Sheet showing the shape of a stone on a particular plane.
Thin section: Polished sliver (usually 30 microns) of stone mounted on a slide for microscopic examination.
Thixatropic: Describing a substance that becomes less viscous when subjected to an applied stress such as shaking or stirring.
Tooling: Surface finish to stone using a chisel.
Tracery: Geometric-based designs made from dressed stone.
Trammel: Instrument to draw curves.
Transom: Horizontal divider of window.
Vault: Arched roof or ceiling.
Verdigris: Green material or staining caused by acidic attack of cuprous materials.
Vernacular: Everyday buildings of a locale.
Voussoir: Single stone of an arch.
Vugs: Small to medium-sized cavities in stone, usually lined with crystals.
Waste cast: The first cast from a mould to check all is working.
WOPC: White Ordinary Portland cement.

INDEX